T0292024

The Evolution of Molecular Biology

The Evolution of Molecular Biology
The Search for the Secrets of Life

Kensal E. van Holde
Oregon State University
Department of Biochemistry and Biophysics
Corvallis, OR, USA

Jordanka Zlatanova
University of Wyoming
Department of Molecular Biology
Laramie, WY, USA

ACADEMIC PRESS

An imprint of Elsevier

Academic Press is an imprint of Elsevier
125 London Wall, London EC2Y 5AS, United Kingdom
525 B Street, Suite 1800, San Diego, CA 92101-4495, United States
50 Hampshire Street, 5th Floor, Cambridge, MA 02139, United States
The Boulevard, Langford Lane, Kidlington, Oxford OX5 1GB, United Kingdom

Notices
Knowledge and best practice in this field are constantly changing. As new research and experience
broaden our understanding, changes in research methods, professional practices, or medical
treatment may become necessary.

Practitioners and researchers must always rely on their own experience and knowledge in
evaluating and using any information, methods, compounds, or experiments described herein.
In using such information or methods they should be mindful of their own safety and the
safety of others, including parties for whom they have a professional responsibility.

To the fullest extent of the law, neither the Publisher nor the authors, contributors, or editors,
assume any liability for any injury and/or damage to persons or property as a matter of products
liability, negligence or otherwise, or from any use or operation of any methods, products,
instructions, or ideas contained in the material herein.

Library of Congress Cataloging-in-Publication Data
A catalog record for this book is available from the Library of Congress

British Library Cataloguing-in-Publication Data
A catalogue record for this book is available from the British Library

ISBN: 978-0-12-812917-3

For information on all Academic Press publications
visit our website at https://www.elsevier.com/books-and-journals

**Working together
to grow libraries in
developing countries**

www.elsevier.com • www.bookaid.org

Publisher: Sara Tenney
Acquisition Editor: Kristi Gomez
Editorial Project Manager: Pat Gonzalez
Production Project Manager: Priya Kumaraguruparan
Cover Designer: Christian Bilbow

Typeset by SPi Global, India

Contents

Preface

Within the past century, a whole new science has arisen, a new way of understanding biology and medicine. The applications of this science, which has come to be called **molecular biology**, pervade every aspect of our lives today and promise even more in the future. Molecular biology has arisen from roots in biochemistry and genetics—has in fact fused these disciplines to provide an understanding of life at a much deeper level than was hitherto possible (Fig. 1.1). As is often the case with new science, unexpected applications have arisen and created whole new industries.

In this book, we will depict the rise and flowering of molecular biology. We will not attempt an exhaustive history of the field, nor of those scientists who built it. Instead, we shall concentrate on the development and flow of ideas. We would like to demonstrate the complexity of science, the sudden breakthroughs following decades of confusion, the frequent blind alleys of misconception that tend to hinder progress. We would like to show how some ideas are slowly crafted by teams of careful and dedicated workers, whereas others arise from individual strokes of genius.

Finally, while this is a book about science, we will try to avoid esoteric knowledge and extensive detail, either about scientific procedures or about the scientists themselves. Nevertheless, there is much about those remarkable men and women who created this field that demands telling, and we shall include biographical material where appropriate.

Chapter 1

Beginnings

PROLOGUE

To this chapter, there is no prologue. It begins with some of the first attempts to explain the world in natural terms. Before, superstition, in a thousand forms, reigned supreme.

SOME ANCIENT INTUITIONS

The basic precept of molecular biology can be stated quite succinctly: all the myriad forms and processes in living things can be explained in terms of atomic and molecular structures and their interactions with one another. Although that level of understanding has not yet been accomplished (and possibly never will be in view of the extreme complexity of life), we have never yet encountered impassible barriers to that quest. It has come close to realization only in the past century (Fig. 1.1). Indeed, the very term "molecular biology" is new. Therefore, it may seem surprising that the basic idea is more than two millennia old. The Greek philosopher Democritus and his colleagues in the 5th century BC proposed a remarkably simple model for the universe. Everything—tables, chairs, the sun, the moon, grass, even human brains and bodies—was proposed to be composed of elementary indivisible particles called **atoms**. They could not, of course, imagine atoms as we visualize them today, but they correctly guessed that different objects and substances were created by differing combinations of atoms (which combinations we call **molecules**). This is the core of modern chemistry and biochemistry. The extrapolation of this idea into biology is the basis for a "molecular biology." This new science is changing our basic understanding of living organisms, whether they are unicellular as bacteria or multicellular as plants and animals. As a consequence of this basic knowledge, the world we live in is changing.

Although it was at odds with every ancient religion or philosophical school and could not be tested by any techniques that would exist for the next two thousand years, the atomic hypothesis retained adherents throughout ancient times. At about the beginning of the Christian era, the Roman poet and philosopher Lucretius composed a remarkable exposition and elaboration of these ideas in a long poem "De rerum natura," usually translated as "On the Nature of Things." Unfortunately, this work vanished for over a thousand years, until a

The Evolution of Molecular Biology. https://doi.org/10.1016/B978-0-12-812917-3.00001-2

FIG. 1.1 A schematic of the history of molecular biology.

copy (Fig. 1.2) was discovered in 1417 by a manuscript hunter in a monastery in central Germany (probably at Fulda). Long before that, during the thousand years of dark ages following the fall of Rome, much of ancient learning had been lost and destroyed, including the original works of Democritus and his school. What little biology the Greeks or Romans had created had degenerated into a chaotic mixture of unrelated observations and stories of fabulous imaginary beasts and plants.

It must not be thought that rediscovery of Lucretius and the atomists led immediately to a rational science. From the fall of Rome (about 500AD) until the renaissance (c. 1500AD) the church ruled scholastic thought, and such ideas were strongly suppressed. When scholars ventured beyond specifically theological matters, they relied upon a few works of philosophers such as Plato and Aristotle that had survived the dark ages and could be (at least partially) reconciled with Christian theology. The figure who stands out, at the very end of this period, as comparable to a modern scientist, is Leonardo da Vinci. Leonardo combined enormous artistic skill with a skeptical, inquiring mind to describe the anatomy of animals, including humans, with accuracy and attention to detail that would not be rivaled for hundreds of years (Fig. 1.3). Elegant and accurate as Leonardo's anatomical studies were they added little to the understanding of mechanism and function. Great as he was, Leonardo was not a modern scientist, in the sense that he did not present hypotheses and test them by experiments. He was a marvelous engineer and a keen observer of nature. In this sense, he was the forerunner of the great naturalists who

FIG. 1.2   A page from a copy of the manuscript "De Rerum Naturae" by Lucretius. This is probably one of a long series of copies, the originals being long lost. *(From https://en.wikipedia. org/wiki/De_rerum_natura).*

would dominate biology in the 18th and 19th centuries. Even when the renaissance opened whole new vistas in astronomy, physics, chemistry, and philosophy, progress in biology was inhibited by two fundamentally erroneous concepts.

SPONTANEOUS GENERATION

The first of these was **spontaneous generation**, the long-held belief that certain simple animals (flies, worms, frogs, etc.) could arise spontaneously from mud, dung, rotten meat, and the like. This was debunked in what was possibly the world's first true biological experiment. The Italian scientist Francesco Redi had observed, like many, that maggots and then flies appeared on rotting meat. But he also noted that flies approaching meat often dropped tiny objects on the meat: he suspected these were eggs. So, in about 1668 he did the following

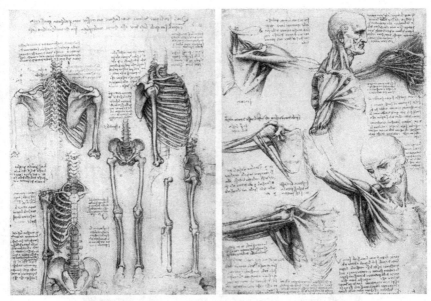

FIG. 1.3 Drawing of human anatomy by Leonardo da Vinci. Such drawings were usually directly from dissections and are usually annotated by the artist. *(From http://www.artcrimearchive. org/article?id=88001).*

experiment: Redi put rotting fish in two bottles, one stoppered and one not. He observed, on repeated trials, that the unstoppered bottle developed maggots, whereas the other did not. Remarkably, Redi still maintained that some other kinds of primitive creatures were generated spontaneously. So did many other biologists, even until the late 19th century, when Louis Pasteur essentially repeated Redi's experiment. Old ideas die hard.

It was, however, another ancient fallacy that most inhibited progress in biology during the renaissance and beyond. This was the doctrine of vitalism, which held that there was some fundamental difference between living and nonliving matter.

VITALISM

Physics and astronomy flourished during the renaissance. Why did not biology? One reason, as we shall see, is that there was an enormous amount of careful data collecting and categorizing that had to be done before the requisite information could be logically ordered. The world of living creatures is incredibly complicated and diverse. But equally inhibiting, in the view of modern scientists, was the influence of a philosophical doctrine termed **vitalism** which asserts that there is a *fundamental* difference in the nature of living versus nonliving matter. Thus, physics and chemistry could never explain life. The basic idea is ancient and seems almost intuitive—living things seem very different from the nonliving. The idea gained power in antiquity from the philosophies of

Plato and Aristotle, with their emphasis on the nonmaterial nature of the "soul." Some of the atomists also believed in the soul, but insisted that it, like everything else, was made from atoms and must obey the same laws.

In the 17th century, at the pinnacle of the renaissance, Galileo and Newton had revolutionized astronomy and physics. Newton's mathematical analysis suggested a wholly physical explanation for how the world works and might have been expected to spell the end of vitalism. However, the influence of the French philosopher Rene Descartes had a profound effect. Descartes introduced a dualistic aspect; the body was material, but inhabited by nonmaterial mind, which could direct its actions. This seemed to allow free will and thus allayed a problem encountered by strict "mechanists." However, it insinuated that there still was something "different" about the behavior of matter in living things. Descartes' compromise may account for the persistence of vitalistic ideas until modern times. Indeed, it has been championed by such authorities as Louis Pasteur, who pronounced in 1858 that fermentation of sugars involved reactions that could only occur in living cells. Curiously, Pasteur derived this conclusion from experiments that definitely ruled out spontaneous generation. In retrospect, vitalism has, in the opinions of many, had a distinctly inhibiting effect on the development of a mechanistic biology.

THE DEMISE OF VITALISM

There are two contenders for the scientific work that definitely turned the tide of scientific thought against the vitalists. The first cites the work of the German chemist, Friedrich Wöhler, who in 1828 accomplished the synthesis of urea from ammonium cyanate. Until then urea, a small molecule containing carbon, oxygen, hydrogen, and nitrogen, had been obtainable only from the urine or kidneys of animals. Ammonium cyanate was recognized by the chemists of the time as an "inorganic" compound, whereas urea was considered "organic." Wöhler's result questioned the distinction between these classes, a part of the vitalist creed. A more devastating blow to vitalism came from the work of Eduard Buchner, who showed in 1897 that in contrast to Pasteur's claim, an extract from broken and dead yeast cells could support fermentation. As biochemistry became a major discipline in the early parts of the 20th century, vitalistic ideas were abandoned by most scientists. Yet even as late as 1913, the eminent British biologist J.S. Haldane questioned whether mechanistic models could ever completely explain life. One may wonder why some ideas, even when discredited, die so hard. Perhaps in this case there is an emotional connection; the thought of a completely mechanistic world is bleak and cold to many.

THE RISE OF MODERN BIOLOGY

All sciences, in their development, seem to pass through a stage of collecting, assembling, and organizing information. Cosmology could not explode until data about thousands of stars and galaxies were accumulated. Chemistry was largely the chaos of alchemy until the chemical elements could be recognized and systematized in the periodic table. Biology, in its formative years, faced a formidable task. There are millions of kinds of organisms, some clearly related, others of no obvious affinities. Organs and body plans may reflect a myriad of life styles.

The biological studies of antiquity did not begin to accomplish such goals. Even at their best, they were sporadic and noncomprehensive anecdotes. In many cases, data about real plants and animals was interspersed with accounts of fantastic creatures, gathered by rumor or hearsay. When the excited minds of the renaissance began to look seriously at biology, the appropriate starting point was to make order out of this chaos—recognizing only what was demonstrably true and placing it in a sensible context. Leonardo da Vinci was the pioneer, examining the growth and forms of plants and animals, generally disregarding the "authority" of the classic writers. His work on anatomy is a marvel in this respect. But the anatomist who probably had the greatest influence during this period was Vesalius, born in Belgium in 1514. He began studies of medicine in Paris, but soon, disillusioned by the uncritical scholasticism, began his own careful studies, including dissections of many animals. In 1543, at the age of 29, he published "De Humani Corporis Fabrica" (The Composition of the Human Body). This massive work, which was accompanied by excellent illustrations, served as a physiology text for generations. Although Vesalius still gave some deference to the classical writers, he found enough to question in them to earn him public condemnation from scholars and clerics of the time.

In terms of taxonomy, the renaissance exhibited only the beginnings, stifled by the still heavy hand of Aristotle. A partial exception was the work of a Swiss, Conrad von Gessner, born in 1516. He published a massive "Historia animalium," which, although still basically Aristotelian in organization, at least excluded many of the more gross errors of the ancients. The birth of a modern taxonomy would have to wait two centuries. In 1735 the great Swedish taxonomist Carolus Linnaeus published "Systema Naturae" which introduced for the first time the binomial system we utilize today—in which each organism is given a **genus** name followed by a descriptive **species** name (e.g., *Homo sapiens*). Over the remainder of the 18th century the process of classification proceeded apace, providing the basis for a systematic biology. Note, however, that this biology, as well as early anatomical studies, was restricted to what the unaided eye could examine. To go deeper into biological structure, the senses must be aided. By 1700, such aid was at hand, in a spectacular fashion.

THE MICROSCOPE OPENS A NEW WORLD

It is often stated that Anton van Leeuwenhoek invented the microscope. This is not strictly true—there were prototype instruments as early as 1609. But it required van Leeuwenhoek's laborious improvements to fashion the instrument (Fig. 1.4) that opened a whole new world of biology, as brought to wide attention in his 1696 book, *Arcana Naturae*. He was able to describe, and even attempt to classify, various bacteria and other one-cell creatures that had never been imagined. In a very short time, a whole new world of biology was opened. On the other hand, Leeuwenhoek held to some old beliefs. Although he observed human sperm in detail, he remained a "spermatist," contending that the whole determination of the being-to-be was held therein.

As microscopes were improved, the "fine structure" of life became apparent. In 1665, the British biologist Robert Hooke first described cells in thin slices of cork, although it is not clear that the generality of this structure was appreciated until later. Indeed, it was not until the 19th century that the detailed structure of the cell came under study. Nevertheless, the impact of the microscope is probably the first example, in the history of biology, when a new instrument reshaped the field. We shall see many more.

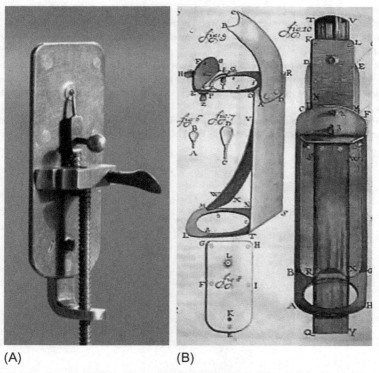

(A)   (B)

FIG. 1.4   Van Leeuwenhoek's microscope. (A) Replica (*from Wikipedia*). (B). Schematic of the microscope as rendered by Henry Baker, naturalist (*from Wikipedia*). Leeuwenhoek's single-lens microscopes used metal frames, holding hand-made lenses. They were relatively small devices, which were used by placing the lens very close in front of the eye. The other side of the microscope had a pin, where the sample was attached. There were also three screws to move the pin and the sample, along three axes: one axis to change the focus, and the two other to move the sample. (*From https:// commons.wikimedia.org/wiki/File:Van_Leeuwenhoek%27s_microscopes_by_Henry_Baker.jpg*).

EPILOGUE

Although the Greek atomists provided a premature insight into a materialistic science, most ancient thought in biology was dominated by careless observation, and fabulous stories. Even the advent of the renaissance added little, for the new thought was at first concentrated on reviving the wisdom of the Greeks and Romans. Unfounded beliefs, including spontaneous generation and vitalism, still held sway. Even careful observation was confined to a few, of whom Leonardo da Vinci was outstanding. Only at the enlightenment, around 1700, were old beliefs challenged, and modern science born.

FURTHER READING

Capra, F., 2007. The science of Leonardo. Inside the Mind of the Great Genius of the Renaissance. Anchor Books, New York, NY. A detailed, beautifully illustrated description of his work in various sciences.

Greenblatt, S., 1994. The Swerve: How the World Became Modern. W. W. Norton & Company, New York, NY. A fascinating account of the rediscovery of Lucretius' masterpiece, and a summary of its contents.

Schrödinger, E., 1944. What is Life? Cambridge University Press, Cambridge, UK. This little book, by an outstanding physicist, inspired many other physicists to enter biology in the postwar period. It is of interest today because of its prescience, and in showing how little we really knew as late as 1944.

Serafini, A., 2001. The Epic History of Biology. Perseus Books Group, New York, NY. Not so strong on the scientific ideas as on the scientists themselves; especially good for the earlier years.

Chapter 2

The Origins of Biochemistry

PROLOGUE

In the late renaissance, around 1700, both chemistry and physiology were emerging as defined sciences. A consequence of their overlap was the definition of a new science, **biochemistry**—the chemistry of living organisms, now largely freed from the constraints of vitalism. We shall see that biochemistry, in turn, fused with genetics to lead to molecular biology. It is the aim of this book to document that remarkable fusion. To do so, we must first describe briefly the backgrounds of genetics and biochemistry, and what each had to contribute. We shall consider biochemistry first, simply because it began earlier. We cannot hope to even summarize this vast field in a few chapters. Rather, we shall trace the understanding of a major class of biochemical substances, the proteins. Proteins are central to almost all biochemical processes, and in particular their interaction with genes is the key to understanding genetics at the molecular level. So, to follow the evolution of molecular biology, it is necessary to understand a bit about proteins.

RECOGNITION OF PROTEINS

It was at the height of the "enlightenment," that period near the close of the 18th century when all of the sciences began to coalesce into their present form. Chemistry, in particular had just seen the massive contributions of men like Priestly, Lavoisier, Cavendish, and Berthelot. In just this period, a few scientists had begun the first tentative studies of substances from living creatures. One group of biologically derived substances, which included albumins from egg and blood serum, gluten from wheat, and the hemoglobin from blood, was recognized by two features—they were rich in nitrogen and they were water-soluble but coagulated by heat. They had been studied briefly in the late 1700s but were only recognized as a distinct class and named "protein" in an 1838 correspondence between the Dutch Gerardus Mulder, and the great Swedish chemist, Jöns Berzelius. We may arbitrarily denote this date as the birth of protein chemistry (Fig. 2.1).

Of course, the chemical techniques of the day could reveal little else about these substances. However, there was one method at which the chemists of that time were superb: quantitative analysis of elemental composition. Their results compare well with what we can do today. Such studies, carried out on the few

The Evolution of Molecular Biology. https://doi.org/10.1016/B978-0-12-812917-3.00002-4

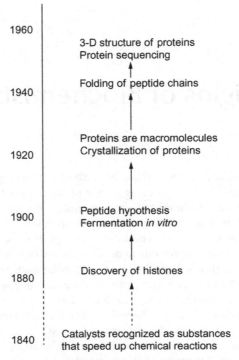

FIG. 2.1 **Time line for the major developments in protein biochemistry during the last two centuries.**

proteins available, revealed a curious fact: There was often one or two elements which were present in very small amounts, compared to the usual C, H, O, and N. Now, the minimum number of atoms that could represent part of a molecule is one, so the protein molecule must be very large to account for such a result. To take an example, hemoglobin from many animals was found to contain one iron atom for a "molecular mass" if about 16,000 (H-units). The idea of such a giant molecule seemed absurd at the time, so the result (which was correct) was largely neglected for almost a century.

SOME PROTEINS ARE CATALYSTS: ENZYMES

The ability of yeast to ferment sugars to alcohol, or the animal stomach to digest foodstuffs, had been realized for millennia, but never understood. But in the early years of the 19th century, the first physiologists had begun to recognize that extracts from some tissues could favor such reactions in vitro (in glass, in the test tube). For example, by 1836 the effect of "gastric juice" on meat was recognized, and in that year Berzelius coined the word **catalyst** to describe a substance that could accelerate a chemical reaction without being modified itself. It must be understood that at the time the nature of the catalytic agents was unknown. Those found in cell extracts were termed "unorganized ferments," not necessarily thought to be connected to an intracellular activity. However, by 1876 enough examples had been studied that the name **"enzyme"** was proposed, and in 1881 the great German physiologist Felix Hoppe-Seyler postulated that these agents were not only present in cells, but that they catalyzed all of the physiological processes. The final clarifying experiment was that of Eduard Buckner, who in 1897 showed that an extract from broken yeast cells could catalyze fermentation.

By this time, most practitioners of the new science of biochemistry believed that enzymes were proteins, apparently on the weak evidence that such activities were often found in the "albuminous," water-soluble fraction of cell homogenates. Unfortunately, further progress was somewhat hindered by a fundamental dispute concerning the nature of proteins themselves. In the early 20th century, the burgeoning of **colloid** chemistry led many scientists to doubt that proteins possessed defined molecular structures. Rather, they might be ill-defined colloidal aggregates of small molecules. This debate was not fully resolved until the 1930s, as we shall explain in a later chapter. It severely hindered attempts to think about protein function at a molecular level.

Actually, solid evidence for a defined molecular structure of at least some proteins had existed as early as 1871, when Wilhelm Preyer published his studies of the formation of crystals of hemoglobin proteins from the blood of a wide variety of animals. It takes molecules, usually identical molecules, to form a crystal. Colloids will not crystallize. Why this evidence was so little noted is difficult to understand, as is the great emphasis in many histories on the crystallization of the enzyme urease by James Batcheller Sumner in 1926. The credit for first crystallizing proteins, and thereby providing evidence that they could not be colloids, should go to Preyer.

WHAT ENZYMES DO, AND WHY IT IS SO IMPORTANT

Enzymes are catalysts; they can speed up certain chemical reactions while remaining unchanged themselves. Most enzymes are specific – they will accept only one (or sometimes two) kinds of molecules (called **substrates**) and accelerate a specific reaction involving these substrates. For example, the hydrolytic splitting of the disaccharide sucrose into the monosaccharides glucose plus fructose is catalyzed by the enzyme **sucrase** (Fig. 2.2), which is specific for this reaction. There are an enormous number of biochemical reactions that are facilitated in this way—examples of enzyme activities recognized early are shown in Table 2.1. It is fair to say that any biochemically important reaction has an enzyme to catalyze it.

It is important to make it clear that enzymes do not drive reactions that would not otherwise proceed—they only make certain reactions go faster. To take an example: we shall see in a later chapter that DNA can be hydrolyzed

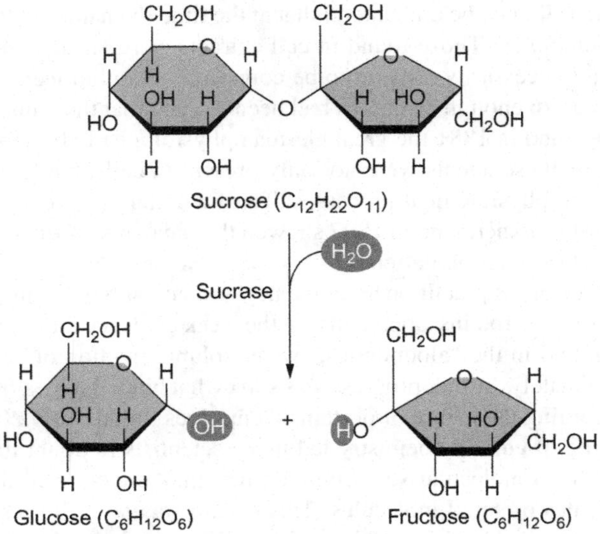

FIG. 2.2   **Hydrolytic splitting of the disaccharide sucrose into the monosaccharides glucose plus fructose, catalyzed by the enzyme sucrase.**

TABLE 2.1 Some Early-Recognized Enzyme Functions

Enzyme	Function	Source	Year
Diastase	Starch hydrolysis	Malt extract	1833
Pepsin	Protein hydrolysis	Gastric juice	1836
Lipase	Fat degradation	Pancreatic juice	1846
Laccase	Oxidation	Latex	1895

(Adapted from Table 15.1 in Tanford, C., Reynolds, J., 2001. Nature's Robots: A History of Proteins. Oxford University Press, Oxford, UK, with permission from Oxford University Press.)

(cleaved by the addition of water). In the presence of an enzyme called a nuclease, this is very fast—a whole long molecule can be broken down to its subunits in seconds. But the uncatalyzed reaction is extremely slow. We can extract DNA molecules from dinosaur bones because the spontaneous DNA breakdown can require ages.

It is this *selective* acceleration of reactions that makes enzymes so important in biochemistry. Consider some foodstuff that is taken into the body. A compound in the food could potentially undergo a myriad of reactions, but there is only *one* reaction pathway that will yield the needed product. Undergoing a successive series of steps, each of which is catalyzed by a specific enzyme, the substrate is processed into the desired product. Much of biochemistry in the late nineteenth and early twentieth centuries was devoted to the unraveling of an amazing network of such paths (for a tiny sample, see Fig. 2.3). It should be noted that the regulation of these metabolic paths can be complex, with enzymes on one path being activated or inhibited by participants in the same or another path. The elucidation of the metabolic landscapes of many organisms was a triumph of careful biochemical studies by a host of scientists. Although our understanding of metabolic pathways is very detailed, the mechanisms by which enzymes function and are regulated continue to be the objects of study.

HOW DO ENZYMES WORK?

There are two basic questions to ask about enzymes: how are they so specific in choice of substrates, and how do they accelerate reactions? The first question was given a reasonable answer in 1897 by the pioneering German biochemist Emil Fischer. Fischer proposed that the surface of the enzyme molecule (whatever that was) could fit the substrate like a lock fits a key (Fig. 2.4A). This can explain specificity but does not satisfy the second question, for a lock does nothing to a key! The problem remained unsolved until 1958, when the American Daniel Koshland proposed the **induced fit** model (Fig. 2.4B). This proposes that the enzyme will fit the substrate only if the latter is distorted into a conformation part-way into the reacted form. This makes it easier for the substrate to go the rest of the way. With modifications, this model is accepted today.

PROTEINS FULFILL MANY ROLES

Enzymes are by no means the whole protein story. There are whole classes of proteins that are involved in maintaining the structures of cells and tissues. The skin, muscle, and nervous tissues each have specific **fibrous proteins** that fulfill their structural functions. There are proteins that act as carriers of small molecules, some of which are signals for metabolic events. There are proteins that regulate gene expression and direct the development of organisms. These are usually present only in the cell nucleus. One class of such proteins, the **histones**, is of interest here because of what it tells about the technical difficulties in the 19th century. To obtain purified nuclei from many tissues was impossible at the time because the crude methods for fragmenting tough materials like muscle or organs would mix nuclear materials with the cell homogenate. A young German scientist, Friedrich Miescher, working in the lab of Hoppe-Seyler, adopted, in 1871, a novel approach. He used, as a source of material, pus from bandages from the local hospital. Such a soft "tissue" allowed gentle cell lysis and isolation of nuclei. Miescher obtained a substance which we now call **chromatin**—the material that makes up chromosomes. It contained a phosphorus-rich component (now called DNA) and a group of proteins, histones. Another student in the lab, Albrecht Kossel, was able to isolate the protein component in 1884, using blood cells from geese. We shall have much more to say about histones in later chapters.

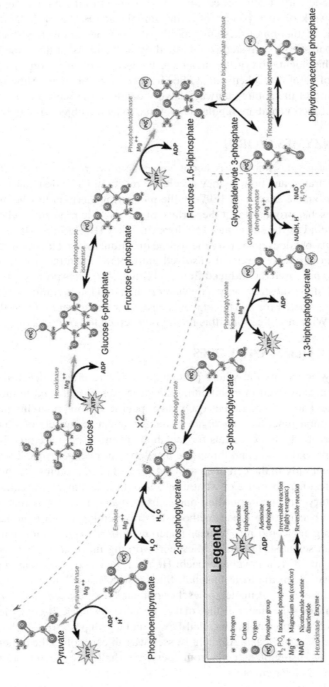

FIG. 2.3   The metabolic pathway for glycolysis, the process that converts the sugar glucose $C_6H_{12}O_6$, into pyruvate, $CH_3COCOO^- + H^+$. The free energy released in this process is stored and used by the cell in the form the high-energy compounds adenosine triphosphate (ATP) and reduced nicotinamide adenine dinucleotide (NADH). *(From commons.wikimedia.org. Author: Yassine Mrabet.)*

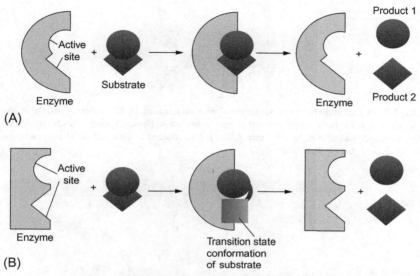

FIG. 2.4   Two models to explain enzyme function. In this specific example, the reaction catalyzed is a cleavage reaction. (A) In the early lock-and-key model, the active site in the enzyme fits snuggly the substrate, as a lock does a key. (B) In the induced-fit model, both the enzyme and the substrate are distorted upon binding. The substrate is forced into a conformation resembling the transition state and the enzyme keeps the substrate under strain, which facilitates the reaction.

The elucidation of these many classes of proteins (including enzymes) and how they are tailored to their diverse functions has been, and continues to be, a vital part of biochemistry. There is overlap here with molecular biology when we begin to consider protein function and mechanisms of action at a molecular level. As we shall see later, it is now becoming possible to study some protein operations at the *single-molecule* level, i.e., one individual molecule at a time.

WHAT ARE PROTEINS MADE OF?

We have been getting ahead of history; it is necessary to step back and ask how the understanding of the nature of proteins developed. The first step was the gradual realization, during the 19th century, that proteins were somehow comprised of alpha amino acids (Fig. 2.5).

These small molecules were always found as products of **hydrolysis** of proteins, either by heating in acid or by digestive enzymes. The general model of an alpha-amino acid is shown in Fig. 2.5; different members of the class are distinguished by the "side group" (R) which, in proteins, may take any of the diverse forms depicted in Fig. 2.6. Note that these provide the protein molecule with a remarkable array of chemical interfaces—acidic or basic, water-avoiding (**hydrophobic**) or water-liking (**hydrophilic**), simple hydrocarbon chains or

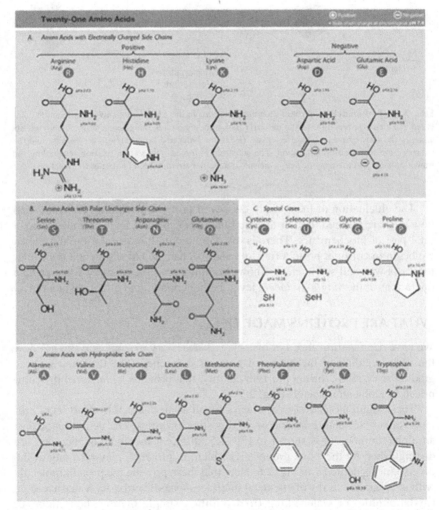

FIG. 2.5 A ball-and-stick representation of an α-amino acid. In these amino acids, the amine group is attached to the α-carbon, the carbon next to the carboxyl group. All amino acids contain this core structure, but differ in the side chain R, that is also attached to the α-carbon. *(From commons. wikimedia.org. Author: Yassine Mrabet.)*

FIG. 2.6 Amino acids found in proteins. The amino acids are grouped according to the chemical properties of their side chains, R. Each amino acid is presented by its full name and its three-letter abbreviation.

rings, etc. This large vocabulary allows proteins to have a multitude of inter-actions, both internal (within the same molecule) and external (between mol-ecules). Basically, this is what enables proteins to do so many things.

EPILOGUE

As the 19th century came to a close, it had become evident that there existed a host of different proteins, adapted to a myriad of biochemical functions. A hint as to their versatility was provided by the recognition that each seemed to have a unique amino acid sequence.

The recognition of the whole armory of amino acids found in proteins began in the early 19th century but required over a century of dedicated chemistry. However, as late as 1900 nobody understood just how a set of such building blocks might be assembled to make a protein. This date marks the true begin-ning of protein chemistry, and a new age of biochemistry. There were, at this point, two great questions to be asked: how were amino acids put together to make a protein, and what was the three-dimensional structure of the resulting product? It is often true in science that progress accelerates when and only when such defined questions can be proposed.

FURTHER READING

Matthews, C.K., van Holde, K.E., Appling, D.R., Anthony-Cahill, S.J., 2013. Biochemistry. fourth ed Pearson, Toronto, Canada. A comprehensive textbook of Biochemistry. Good for details, but not much on history.

Tanford, C., Reynolds, J., 2001. Nature's Robots: A History of Proteins. Oxford University Press, Oxford, U.K. A very comprehensive and readable history of protein chemistry; probably the best source to date.

Chapter 3

The Chemical Structure of Proteins

PROLOGUE

By the beginning of the 20th century, proteins had become recognized as important, ubiquitous, but still mysterious components of the cell. Their chemical composition was known to be rich in nitrogen. If they were treated with hot acid, they yielded a mixture of small molecules known collectively as amino acids. Those amino acids that had been isolated and studied all had the general structure shown in Fig. 2.5. They differed only in the group designated R, which exhibits some 20 different forms (Fig. 2.6), giving proteins the possibility of great complexity and versatility. How were these amino acids assembled to make proteins? What did this mean in terms of biological function?

THE PEPTIDE HYPOTHESIS

How the amino acids might fit together to produce proteins was completely unknown in 1900. Then, at a meeting in Carlsbad in 1902 two eminent scientists, Franz Hofmeister and Edwin Fischer proposed that two such molecules might join, by elimination of a water molecule between the **amino group** ($-NH_2$) of one and the **carboxyl group** ($-COOH$) of another to form what Fischer termed a "peptide" (Fig. 3.1A). Indeed, there was no reason for the process to be limited to two amino acids; in fact, Fischer succeeded in fashioning **polypeptide** chains up to 18 units in length. Such chain-like molecules might contain any of the amino acids, linked through "**peptide bonds**." Each chain would have an unreacted "**N-terminal end**" and an unreacted "**C-terminal end**" (Fig. 3.1B). But did such structures exist in proteins? The evidence at the time was slim indeed, resting solely in the fact that Hofmeister was able to isolate a dipeptide of the amino acids glycine and alanine from breakdown of fibroin, the major protein component in silk. Fibroin was, in fact, a lucky choice, for it is one of the few proteins in which repetitions of dipeptides are common.

Fischer and Hofmeister had been working independently and apparently had not known of each other's results. They approached the problem in entirely different ways: Fischer a superb "bench chemist" had succeeded in making small peptides by dehydration reactions, whereas the more cerebral Hofmeister had

The Evolution of Molecular Biology. https://doi.org/10.1016/B978-0-12-812917-3.00003-6

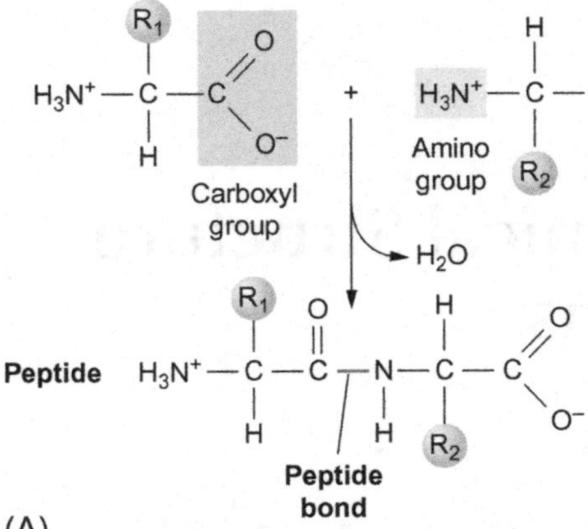

(A)

(B) H₂N-Gly Ser Gly Ala Gly Ala-COOH

FIG. 3.1 **Amino acids and the peptide bond.** (A) Chemical structure of α-amino acids. R denotes a chemical group that is unique to each amino acid. The asymmetric C atom satisfies its four valencies by the attachment of four different groups. A peptide bond is formed by elimination of a molecule of water from between two amino acids. The resulting compound is termed a "peptide." (B) An example of a small peptide of physiological importance. This is human somatostatin, a neurotransmitter. By convention, peptide sequences are written with the N-terminus to the left. You can identify the amino acids from Fig. 2.6.

deduced the peptide bond as the only logical way that amino acids could link together to form peptides. His approach resembles that of Linus Pauling, who generations later deduced the logical ways in which polypeptide chains could fold (see Chapter 4).

Needless to say, the implication that proteins could be long polypeptide molecules was not readily accepted by the chemists of the day. The whole concept of giant molecules that could be formed by addition of similar units (which we now call **polymers**) seemed just too untidy. Even as late as 1924, the German chemist Herman Staudinger was derided for such proposals. Some kind of demonstration with real proteins was needed. Examples like the low iron content of hemoglobin (Chapter 2) seemed too abstract. How could one directly measure the molecular weight of a large molecule?

COLLOID OR MACROMOLECULE?

Were proteins even molecules at all? The question stemmed from the dominance of the idea of colloidal aggregates of small molecules. The possibility that proteins were such colloids was put forward as an explanation for some peculiar behavior of protein solutions, observed in many laboratories. The very low osmotic pressure produced by putting a protein in solution seemed to point to absurdly high molecular weights for many protein molecules. It was easier for chemists of the age to propose that these were not "true" solutions of molecules, but some special colloidal state of matter. Further, as late as 1924, it was proposed that even the basic peptide structures described by Hofmeister and Fischer were irrelevant—maybe proteins were colloidal aggregates of an entirely different chemical structure, and the amino acids produced on protein breakdown were artifacts of the process. Although discredited by many, this wholly erroneous model persisted for at least a decade. Decades after Fischer and Hofmeister, the molecular structure of one of the basic components of life remained contentious. Colloidal aggregates did exist (as in soaps and detergents, for example); was this how proteins were constructed?

SOME UNEXPECTED RESULTS

The problem was resolved from an unexpected source. Theodor ("The") Svedberg was a Swedish chemist at the University of Uppsala, interested in the size distributions of particles in colloidal suspensions. To study this, he would allow a sample to sediment in a glass tube; the heavier particles would sink more rapidly, the smaller ones would lag behind. If the suspension consisted of very small particles, sedimentation under gravity became very slow and unstable. Therefore, Svedberg hit upon the idea of using centrifugal force, rather than gravity, to sediment colloids. From an old cream separator, he constructed a contraption in which samples of solutions were held between glass plates in a spinning rotor. This allowed Svedberg to observe the sedimentation as it was taking place, provided that the sedimenting material had different light absorbance than the solvent. Svedberg called this instrument an "ultracentrifuge." The first models were so successful that Svedberg built more powerful machines, capable of sedimenting large molecules (for more information on the ultracentrifuge, see section "Sedimentation" in Box 3.1).

About 1925, Svedberg and his colleague Robin Fahraeus decided to apply their new technique to a protein. They chose horse hemoglobin for a first trial, partly because it was easy to obtain and purify, and because of its distinctive color. They used a variety of the technique called sedimentation equilibrium, in which, after a time of spinning, sedimentation of the particles is balanced by back diffusion. The situation is similar to that in the atmosphere, where heavier molecules like CO_2 will concentrate at lower levels, and light molecules like hydrogen will distribute more evenly. Recall that Svedberg's laboratory focused on colloids; they likely expected typical colloidal behavior. If hemoglobin particles were random colloidal aggregates, their distribution at equilibrium should show smaller, lighter hemoglobin particles concentrated toward the top of the solution and larger, heavier ones toward the bottom. But this is not what was found; instead, the particle weight observed was approximately uniform over the whole solution column! All of the hemoglobin particles were of the same size!

BOX 3.1   Separation Methods

Modern biochemistry depends strongly on a number of methods to separate cell constituents, different macromolecules, and macromolecules from small molecules. To study a particular protein, for example, it is necessary first to isolate the cellular structure in which this protein can be found (nucleus, cytoplasm, etc.). Then the protein must be purified from contaminating proteins and small molecules. Often, separation methods can also be used to obtain physical parameters of the molecule—it size, shape, molecular mass, electric charge, etc. It was the development and sophistication of such methods over the past 200 years that has allowed our present detailed knowledge of proteins, nucleic acids, and other molecules of life. We cannot describe these methods in detail here or do we need to. We will, however, in this and the following box briefly describe certain methods that have been essential for the continued development of molecular biology.

Sedimentation

For a long time, simple bench-top centrifuges have been a component of every biochemical laboratory. They can hasten precipitation and separate cell components. They can isolate fats, because fatty materials float on aqueous solution—the basis for the long-established method to determine butter fat in milk. An analytical ultracentrifuge is a very different apparatus. The primary difference is not only in rotor speed (which may be very high, up to 100,000 RPM, rotations per minute) but also in the fact that such an instrument is equipped with an optical system that can follow the sedimentation of molecules in "cells" in the rotor. Modern ultracentrifuges are all derived from Svedberg's original design but are highly sophisticated in automated operation and data retrieval and processing.

Basically, such an **analytical ultracentrifuge** is devoted to measurement of molecular parameters, not to preparative separations as described earlier. There are two basic kinds of analysis: sedimentation velocity and sedimentation equilibrium. The first follows the course of sedimentation of dissolved molecules from the solution meniscus to the bottom of the cell. This may reveal the presence in the sample of more than one macromolecular component, and the velocity of each can provide quantitative information about molecular mass, size, and shape. For such analysis, high rotor speeds will give the best resolution.

The sedimentation equilibrium method uses lower rotor speeds to produce a gentle sedimentation, which is ultimately balanced by back-diffusion of the molecules, yielding an equilibrium gradient of concentration. Analysis of this gives an unequivocal value for the molecular mass. This was the technique first used by Svedberg and Fahraeus to determine the molecular mass of hemoglobin. Homogeneity of hemoglobin was later confirmed by sedimentation velocity experiments.

Chromatography

The term **chromatography** covers a multitude of variations on a common theme. A mixture of substances in solution are absorbed on some material and then eluted from it (literally "washed out") by the passage of a solvent in which the substances bind less strongly to the absorbent. Under such conditions, different absorbed components in the mix will travel at different rates in a column of the absorbent and can be detected and identified as they emerge from the column. This method is referred to as **column chromatography.**

BOX 3.1   Separation Methods—Cont'd

A major and important modification of this technique is called **paper chromatography**; here the sample to be analyzed is simply added as a drop on a sheet of cellulose paper and is eluted by passing the eluting solvent down the paper; spots corresponding to different components will move at different rates. They can then be identified chemically, or by radioactivity labeling. In a powerful, and widely used variant, the sample is spotted near one corner of the paper and eluted successively by two solvent systems. This gives a two-dimensional map of components (see Fig. 3.2).

Electrophoresis

Most macromolecules, including proteins, carry a net electrical charge. In proteins, this is because positively charged side chains, like lysine or arginine, are not present in equal quantities as compared to negatively charged groups like aspartic and glutamic acids (see Fig. 2.6). Furthermore, because these kinds of groups are weak bases and acids, respectively, the net charge (+, 0, or −) will depend on the pH of the solution. The consequence of all of this is the fact that a method that separates molecules according to charge can be very powerful and versatile.

Such a method is **electrophoresis**, which exists in a myriad of variants. They all depend upon the fact that an electric field, maintained by a potential difference between two electrodes, will exert a force on charged particles in an intervening solution, causing them to move. Each moves in a direction determined by the net charge: positive particles toward the negative electrode, negatives vice-versa, and uncharged move not at all. For most protein molecules, there will be some pH values at which negative and positive charges just balance; this is called the **isoelectric point**. The velocity of motion, and hence the distance each molecule is moved in a given time, depends both on the magnitude of the charge and the resistance the medium presents to particle movement.

The most convenient way to perform electrophoresis is on a supporting medium. This can be paper, utilized in much the same way as in the paper chromatography described above, only now the wetted paper strip on which the sample is dotted stretches between two electrode compartments. The resulting distribution of spots is then revealed by some method of staining.

By far the most popular and versatile method in the last several decades is **gel electrophoresis**. In this technique, the supporting medium is a gel (originally agarose, a natural polysaccharide, but more recently the synthetic polymer, polyacrylamide). A common apparatus is depicted in Figure 1. A slab of gel, prepared with the desired **buffer** solution is cast between glass plates; the slab contains "wells" at the top in which samples can be placed. The top and bottom electrodes are then connected to a regulated power source. After small "marker" molecules, colored to allow observation of their movement during the electrophoretic run, have migrated to the bottom of the gel, the slab is taken out and stained, or otherwise treated to reveal the bands in each lane. We will see many applications of variants of gel electrophoresis in subsequent chapters.

Continued

BOX 3.1   Separation Methods—Cont'd

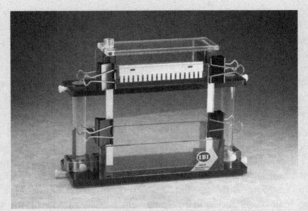

FIGURE 1.   Typical apparatus used for vertical polyacrylamide gel electrophoresis. The gel is cast between the two glass plates, which are separated by thin plastic spacers and held together by the paper clips. The "comb" at the top will be removed, leaving wells into which the samples will be loaded.

In the detailed study of proteins that emerged in the second half of the 20th century, two variants of gel electrophoresis have played a prominent role—**isoelectric focusing** and **sodium dodecyl sulfate (SDS)-gel electrophoresis**. The former can be understood by recalling that at the isoelectric point a protein cannot respond to an electric field. If the gel slab is prepared with buffers giving a gradient of pH, and a sample of a protein mix loaded in a well, each protein will migrate toward the point of its isoelectric point and stop there. Thus, the proteins will be sorted according to their charge distribution.

It was long realized that it would be convenient to have an electrophoretic method to estimate protein molecular mass. However, the migration of proteins in gels depends on a number of factors besides mass—electric charge, size, and shape of the molecule, and sieving effect of the gel. It turns out that many of these complications can be sidestepped by simply including in the sample and the gel a detergent like **SDS**. Remarkably, in such conditions, the rate of electrophoresis is primarily determined by the mass of the protein. Reasons are complicated, but essentially it seems that the detergent coats and unfolds a protein molecule, making length (and hence mass) the determining factor in mobility. Such gels need to be calibrated by sets of proteins of known mass but provide a quick and easy way to estimate the mass of a protein.

If one wishes to quickly characterize a complex mix of proteins—like all those in a particular bacterium species, **two-dimensional electrophoresis** is very powerful (Figure 2). Here, isoelectric focusing is carried out in one dimension, and SDS gel electrophoresis in the perpendicular dimension. It is fair to say that electrophoresis, in its many forms, has played a critical role in protein biochemistry. We shall see later that it has been of equal or greater importance in molecular genetics and molecular biology.

BOX 3.1   Separation Methods—Cont'd

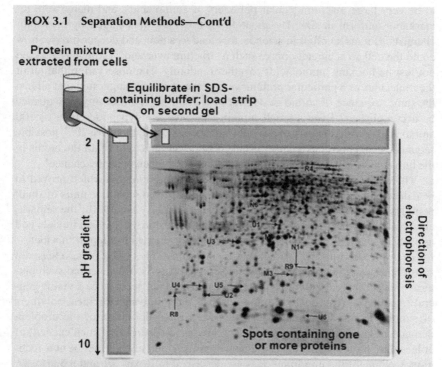

FIGURE 2. Two-dimensional gel electrophoresis for the separation of complex protein mixtures. The isoelectric-focusing first dimension gel is not stained for proteins because the stain interferes with the second-dimension gel electrophoresis. *(From Jagadish, S.V.K., Muthurajan, R., Oane, R., Wheeler, T.R., Heuer, S., Bennett, J., Craufurd, P.Q., 2010. Physiological and proteomic approaches to address heat tolerance during anthesis in rice (Oryza sativa L.). J. Exp. Bot. 61, 143–156, Fig. 4a, with permission from Oxford University Press.)*

This was a remarkable and unexpected result. It meant that the hemoglobin particles were not likely to be colloidal aggregates. They must be precisely defined molecules. Furthermore, the experiment showed them to be big molecules, with mass about 68,000 times that of a hydrogen atom. This is remarkably close to the value of 64,650 known today from its exact amino acid sequence. It was a value surprisingly larger than for any other molecule known at the time. The studies on hemoglobin were soon repeated with many other proteins, with the same kind of results. Svedberg was awarded the Nobel Prize in Chemistry in 1926 for his ultracentrifuge studies.

PROTEINS AS HOMOGENEOUS POLYPEPTIDES

Svedberg's result, together with the Fischer-Hofmeister hypothesis, provided for the first time a working model for protein structure: proteins were now recognized to be very large macromolecular polymers of amino acids, and many were re-markably uniform in size. By about 1935, this view was shared by most bio-chemists. But, as so often in science, this lead to a new and deeper puzzle: how could the cell so accurately make such a structure over and over again? It could not just be hooking amino acids together randomly. The observation that all of the molecules of a particular protein were of the same size suggested that all had the same sequence of amino acids. But how could such uniformity in sequence be accomplished? Even a small protein can have a polypeptide chain of 100 amino acids. With 20 kinds to choose from at each point, there are 20^{100} possible sequences; an enormous number—far greater than the count of all the atoms in the universe. A cell could *never* make the same sequence twice by chance!

There was one seemingly easy solution to the dilemma—which proved to be a dead end. Suppose proteins were put together from repetitive units of small peptides, using specific enzymes to add each kind? By the 1930s, the remark-able versatility of enzymes was known to all biochemists, so such models had appeal. There were even experimental results that seemed to support this idea—for example, fibroin, the protein that mostly constitutes silk fibers, released a substantial fraction of repetitive peptides on hydrolysis. However, no such pat-terns were observed with most proteins, so this did not seem to be a viable gen-eral hypothesis. (Curiously, it turns out that many **polysaccharides**, polymers of simple sugars, are made in just this way.) Final definition of the problem demanded determination of the unique amino acid sequence (if such existed) of at least one protein. That needed a new approach, and a whole set of new tech-niques. Some of the important ones are described in Boxes 3.1 and 3.2.

BOX 3.2   Immunological Methods

Another important way of identifying or separating proteins is via **immunological reactions**. To explain how this can be, we need a few words about immunology. When a higher organism is under attack by viruses, bacteria, or other extraneous substances, it responds by producing protein molecules called immunoglobulins or antibodies that bind to and neutralize the invading agents. The antibodies are actually elicited by a particular substance or a portion of a molecule, called an an-tigenic determinant (or epitope). Large biological polymers usually contain several such portions, and the immunological response results in the production of numer-ous diverse antibody molecules, each of which recognizes, binds to, and neutral-izes a specific epitope. Thus, antibodies are highly specific to individual antigens and bind to them with very high affinity. These properties of antibodies make them a valuable tool in both basic research and clinical practice.

Antibodies are used in a variety of ways and for different purposes. The enzyme-linked immunosorbent assay or ELISA (Figure 1A) and Western blot (Figure 1B) techniques use antibodies to detect the presence of a particular antigen in a sam-ple. In addition, antibodies are used to localize a particular antigen in cytological preparations.

BOX 3.2 Immunological Methods—Cont'd

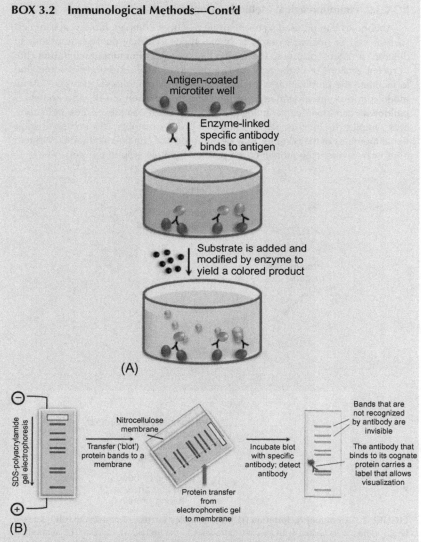

FIGURE 1. Immunological methods. (A) ELISA (enzyme-linked immunosorbent assay). The assay is performed on microtiter plastic plates that contain 96 small wells so that 96 reactions can be carried out simultaneously. The readout of the assay is visible color which is automatically quantified by special reader instruments. The schematic presents only the simplest version of the assay; more sophisticated and sensitive techniques are in use too. (B) Immunoblotting (more commonly known as Western blotting). The name Western blot mimics the name Southern blot, a technique for transfer of nucleic acids from electrophoretic gels to membranes, introduced by Edwin Southern and named after him.

Continued

BOX 3.2   Immunological Methods—Cont'd

Antibodies can be used to purify proteins out of complex biological samples, such as cell or nuclear extracts or bodily fluids that may contain thousands of different protein molecules. The method is known as **immunoprecipitation (IP)**. In recent versions of IP, the antibody is coupled to a solid substrate of some kind to facilitate the purification of the antigen/antibody complex (Figure 2). A useful modification of the IP technique, **coimmunoprecipitation (co-IP,** also known as **pulldown)**, can identify interacting proteins or protein complexes present in complex samples: by immunoprecipitating one protein member of a complex, additional members of the complex may be captured and can be identified by Western blots or by sequencing protein bands purified from electrophoretic gels.

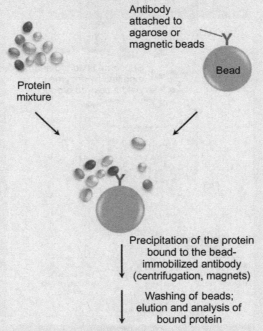

FIGURE 2.   Immunoprecipitation (IP). The complex mixture of proteins in cell lysates or bodily fluids is incubated with specific antibodies to the protein of interest. The antibody is attached (immobilized) on some solid support, usually resin or magnetic beads. The protein of interest interacts with the antibody, and the entire bead/antibody/antigen system can be conveniently separated from the unreacted proteins. This is done by simple low-speed centrifugation in the case of agarose beads, or by just applying magnets to the tube wall in the case of magnetic beads. The precipitated complex is washed with buffer a couple of times to remove any contaminating proteins and the proteins attached to the antibody are eluted and analyzed.

FRED SANGER AND THE SEQUENCE OF INSULIN

In 1943, a young biochemist named Fredrick Sanger who had just obtained his PhD at Cambridge University began postdoctoral research to "investigate the structure of insulin." More specifically, he was assigned by his supervisor to devise a method to determine which amino acid residue lay at the N-terminal end of a protein. Sanger did a great deal more than that, becoming one of the two individuals ever to win two Nobel Prizes.

Insulin was an obvious choice because it was known to be one of the smaller protein molecules, and it was available in quite pure form, as it was already used in the treatment of diabetes. Sanger soon solved his initial problem; he found the amino acids which lay at the N-terminal tails of the two peptide chains in insulin, using a colored compound that selectively reacted to those sites. Then, completely digesting the protein in acid, he could pick out the colored terminal amino acids from all the rest by a new technique called chromatography (Fig. 3.2, see also section "Chromatography" in Box 3.1). There turned out to be two polypeptide chains in the molecule. Sanger soon realized that one might determine the internal sequence of each chain, by using mild acid treatment and enzyme digestion to make many overlapping peptides (Fig. 3.3), which allowed him to deduce the entire sequence of both chains (Fig. 3.4). This required nearly a decade of laborious, careful work, much of which was carried out by Sanger himself.

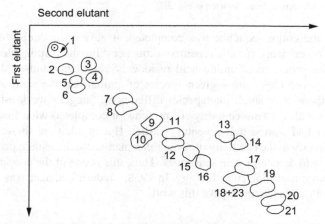

FIG. 3.2 Paper chromatography to separate amino acids or peptides. A drop of the mixture to be analyzed was placed on one corner of a piece of filter paper. The paper was dried and then a solvent mixture is allowed to pass through the paper in one direction. After the different amino acids were carried to different distances, the paper was redried and then washed with a different solvent mixture in the perpendicular direction. This gives a two-dimensional "fingerprint" of the mixture. If one of the amino acids was color labeled, its identity could be seen immediately. *(From Sanger, F.,Thompson, E.O., 1953. The amino-acid sequence in the glycyl chain of insulin. I. The identification of lower peptides from partial hydrolysates. Biochem. J. 53, 353–366, Fig. 3, with permission from Portland Press.)*

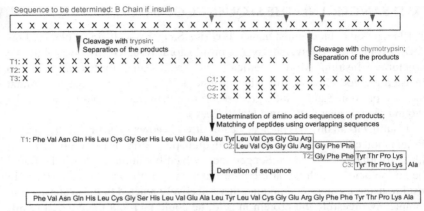

Sequence to be determined: B Chain if insulin

FIG. 3.3   The use of overlapping peptides to sequence polypeptides. *(Based on Mathews, C. K., et al., 2013. Biochemistry. Pearson, Toronto. Fig. 5B.3, with permission from Pearson.)*

A Chain
Gly Ile Val Glu Gln Cys Cys Thr Ser Ile Cys Ser Leu Tyr Gln Leu Glu Asn Tyr Cys Asn
B Chain
Phe Val Asn Gln His Leu Cys Gly Ser His Leu Val Glu Ala Leu Tyr Leu Val Cys Gly Glu Arg Gly Phe Phe Tyr Thr Pro Lys Thr

FIG. 3.4   The amino acid sequence of bovine insulin. The two chains are held together by disulfide (S–S) bonds. Sanger had to break these, separate the A and B chains, and then determine the sequence of each separately, as depicted in Fig. 3.3.

When the entire sequence was completed, it gave an unequivocal result: there was no evidence for any repetitive sequences; bovine insulin was defined as a unique sequence of amino acid residues. Subsequent studies show that all insulin molecules from a given species of animal had the same sequence although there were small interspecies differences. Sanger's work established once and for all the "molecularity" of proteins; it completed what Fischer and Hofmeister had started half a century earlier. But in addition, these findings pointed directly to the necessity of some fundamental mechanism to provide the information to dictate protein sequences. Thus, this is one of the crucial studies in the evolution of molecular biology. In 1958, Frederick Sanger was awarded the Nobel Prize in Chemistry for this work.

EPILOGUE

By about 1950, the basic principles of the chemical structure of proteins were firmly established. Proteins were not colloidal aggregates but were composed of long polypeptide chains and at least most were homogeneous in size and sequence. Sanger's work had driven the last nails in the coffin of many erroneous ideas. Still, two major questions remained: How was the amino acid sequence determined in the cell? What were the three-dimensional physical structures of protein molecules and what determined these?

FURTHER READING

Kessel, A., Ben-Tal, N., 2010. Introduction to Proteins: Structure, Function, and Motion. CRC Press, Boca Raton, FL. Unified, in-depth, contemporary treatment of the relationship between the structure, dynamics, and function of proteins.

Petsko, G.A., Ringe, D., 2003. Protein Structure and Function. New Science Press, London, UK. Introduction to the general principles of protein structure, folding, and function. Also, provides the conceptual basis of inferring structure and function from genomic sequence.

Tanford, C., Reynolds, J., 2001. Nature's Robots: A History of Proteins. Oxford University Press, Oxford, UK. A very comprehensive and readable history of protein chemistry; probably the best source to date.

Whitford, D., 2013. Proteins: Structure and Function. Wiley, Hoboken, NJ. Comprehensive treatise of protein structure from chemical and physical points of view, highlighting the connection between structure and function.

Zlatanova, J., van Holde, K.E., 2016. Molecular Biology. Structure and Dynamics of Genomes and Proteomes. Garland Science. Taylor & Francis Group, New York, NY. Much more detail on proteins/nucleic acids and methods used for their study will be found in this comprehensive textbook.

Classic Research Papers

Edman, P., 1949. A method for the determination of amino acid sequence in peptides. Arch. Biochem. 22, 475. Description of the method that became known as Edman degradation for peptide sequencing. The amino-terminal residue is labeled, cleaved from the rest of the peptide, and identified.

Ingram, V.M., 1958. Abnormal human haemoglobins. I. The comparison of normal human and sickle-cell haemoglobins by fingerprinting. Biochim. Biophys. Acta 28, 539–545. Reports the detection of differences between normal and abnormal haemoglobins by using the method of fingerprinting.

Sanger, F., Thompson, E.O., 1953a. The amino-acid sequence in the glycyl chain of insulin. I. The identification of lower peptides from partial hydrolysates. Biochem. J. 53, 353–366. This and the next paper describe the methods and results of sequencing insulin.

Sanger, F., Thompson, E.O., 1953b. The amino-acid sequence in the glycyl chain of insulin. II. The investigation of peptides from enzymic hydrolysates. Biochem. J. 53, 366–374. This and the previous paper describe the methods and results of sequencing insulin.

Chapter 4

Proteins in Three Dimensions

PROLOGUE

Before the middle of the 20th century, it was generally agreed among biochemists that protein molecules were composed of long polypeptide chains. But this still left open the question as to how they were internally structured. The backbone of a polypeptide contains periodic sites of flexibility—bonds about which swiveling is possible. That meant that the chain could potentially take an almost infinite number of spatial conformations, ranging from full extension to condensation into a globule. To understand protein function required that protein **conformation** be understood.

FIBERS

Two lines of research, both in the 1930s, helped define the limits of protein structure. The first was the pioneering work by a British scientist at Leeds University, William Thomas Astbury, who applied the relatively new technique of X-ray diffraction (Box 4.1) to fibrous structures like wool, silk, and hair. In the hair protein, **keratin**, Astbury was able to distinguish two periodic conformations; one (called alpha) was found only in unstretched fibers, and the second (called beta) only in wet, stretched keratin fibers and in silk. The periodicity of beta keratin was that expected for a stretched polypeptide chain with extended amino acids; alpha keratin had a shorter structural periodicity that had to indicate some kind of a folded chain (Fig. 4.1). These results, although they had little further impact at the time, were to be a crucial importance a decade later, when they led Linus Pauling to consider the general problem of protein chain folding.

Pauling and his coworkers approached the problem of folding from a theoretical point of view. They asked; what do we know about amino acids that would constrain the way in which they can fold into fibers? They had available the incomplete X-ray diffraction result of Astbury and a few other workers. They had some structural information from studies of amino acids and small peptides. And Pauling himself possessed a deep understanding of chemical bonding; he was the author of the seminal book on that topic, "The Nature of

The Evolution of Molecular Biology. https://doi.org/10.1016/B978-0-12-812917-3.00004-8

BOX 4.1 How to Determine Protein Structure

Determining the three-dimensional structure of a protein is a formidable task. The method most commonly used is **X-ray diffraction**. This is a sophisticated physical technique; we cannot hope to present an in-depth treatment in a few pages. However, it is possible to show the basics and point out the demanding problems that made its application to proteins so difficult.

The Principle of Diffraction

To begin with the simplest possible example, consider a row of regularly spaced atoms as in Figure 1, with a beam of radiation being passed through it. Each atom will scatter some of the radiation in all directions. But in most directions the scattered beams from different atoms will be out of phase with one another, and cancel each other out. Only in *certain* directions will the path difference (PD) in the waves scattered from successive atoms equal the wavelength of the radiation, or a multiple thereof. In those and only those directions the scattering from the atoms in the row will add up, and be registered on a photographic plate. It is common to visualize the radiation as "reflected" from the array of atoms. The angle (θ, theta) of reflection at which waves are reinforcing (see Figure 1) can be deduced by simple geometry to be given by

$$\sin(\theta) = n(\lambda)/2d$$

where n is an integer, λ (lambda) is the wave length of the radiation, and d is the spacing between scatterers (see Figure 1). Thus, by measuring the values of θ for diffraction, we can measure d, the spacing in this one-dimensional crystal. This is the fundamental basis for al diffraction studies. Note that small values of d correspond to large values of θ; the observed diffraction pattern spacings are always *inversely* proportional to structural spacings in the sample.

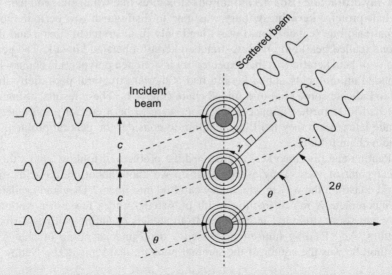

FIGURE 1. Scattering from a row of scatterers. When the angle theta (θ) is just so that the path difference from rays from adjacent scatterers just equals the wavelength or a multiple thereof, there is reinforcement. Otherwise, there will be interference.

BOX 4.1   How to Determine Protein Structure—Cont'd

Diffraction from Fibers

Note that Figure 1 represents an oversimplified model of a polymer fiber. Any real fiber studied will have finite thickness, within which fibrous molecules like keratin or DNA are more-or-less aligned in parallel. We shall expect to see a diffraction pattern with multiple spots corresponding to regular spacings along the fiber. If there is regular structure in how the molecules pack side-by-side, this will give rise to a pattern perpendicular to the fiber axis.

A most important example arises when the fibrous molecules are helical. In this case the spots form a very typical X pattern, with spacings inversely proportional to the **pitch** of the helix (the distance at which it makes a complete turn). As we shall see in Chapter 6, this feature was very important in the study of DNA fiber structure.

Diffraction from Crystals

A molecular crystal is a regular three-dimensional array of molecules, such as sche-matized in Figure 2. Here we depict an imaginary picture of a crystal of a globular protein. "Regularity" means that the molecules are arranged in a three-dimensional lattice, indicated here by the heavy lines. These lines define the **unit cell** of the lat-tice—that unit of volume which repeats over and over to define the whole crystal.

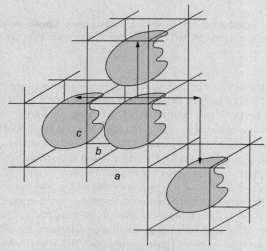

FIGURE 2.   Schematic of a molecular crystal. The "unit cell", the element of volume which by repetition can generate the whole crystal, is indicated by heavy lines. In the simple example here, one globular protein molecule (irregular shaded object) occupies each cell in the lattice.

Now suppose we place a crystal in a beam of X-rays. We choose X-rays because their wave length is about the same as the details we want to study in the crystal. For simplicity, suppose first that the molecule is very simple—a diatomic molecule like $C=O$. In Figure 3 we have depicted these atoms as black

Continued

BOX 4.1 How to Determine Protein Structure—Cont'd

and white. We assume we are looking down parallel planes of atoms edge on, such as –B-B-B-B– and –W-W-W-W–, and that we are seeing the scattering of the X-ray beam from the atoms in these planes. Just as we saw above, at certain angles, indicated by theta, the path difference for rays from successive layers of black atoms will lead to reinforcement. That occurs because the path difference is just equal to the wave length or a multiple thereof. This is the so-called **Bragg angle**, named after a pioneer in X-ray diffraction, Sir Lawrence Bragg. At other angles, there is destructive interference. The angle θ obeys the simple equation we gave above.

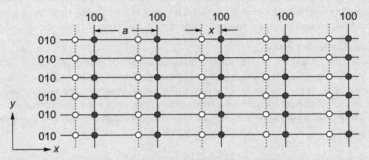

FIGURE 3. A hypothetical crystal lattice containing a simple molecule (C=O). Note that a lattice drawn through the O atoms is identical in spacings to that through the C atoms.

Looking at diffraction from different orientations (as we rotate the crystal in the beam) will yield a pattern of spots on a photographic plate (Figure 4). This can tell us the unit cell type and dimensions but it's not enough to determine the structure of the molecule in each cell. How far is C from O, for example? But there is more information here. Note that the planes of O atoms have exactly the same spacing as planes of C atoms. Therefore, these must contribute to the same spots on the pattern, *but somewhat out of phase with the contribution from the C atoms*. This phase difference will influence the intensity of the corresponding diffraction spots. The effect will be different for other planes in the crystal. Thus, there is information in the intensities of spots in the diffraction pattern concerning the contents of the unit cell. The problem is to determine the relative phases of different spots so as to extract that information. This is the "phase problem" in a vastly simplified presentation. The great contribution of Perutz and his co-workers was to realize that replacing one or more atoms in a complex molecule with a heavier atom would perturb the intensities in ways that could be used to measure the position of other atoms (or more precisely, regions of high electron density). Thus, a picture of the molecule can be constructed.

This is no easy task. Just obtaining good crystals (even tiny ones) from protein solutions is more an art than a science. Then, repeating the process with crystals containing one or more kinds of heavy metal substitutions is further complicated

BOX 4.1   How to Determine Protein Structure—Cont'd

FIGURE 4.   An X-ray diffraction pattern from the protein sperm whale myoglobin.
The pattern is centered about the point where the undiffracted X-ray beam would strike the photographic plate. The resolution that can be obtained in the structure depends on how far out from the center spots are utilized. *(From Voet, D., Voet, J.G., Pratt, C.W., 2013. Fundamentals of Biochemistry: Life at the Molecular Levels, fourth ed. Wiley, Hoboken, NJ, Fig. 6.21, with permission from Wiley.)*

if the metals significantly perturb the structure. No wonder that it took a decade of dedicated work to solve the problem!

Drenth, J., 1994. Principles of Protein X-ray Crystallography. Springer Verlag, New York. Comprehensive, difficult, and reasonably up-to-date.

van Holde, K.E., Johnson, W.C., Ho, P.S., 2006. Principles of Physical Biochemistry. Prentice-Hall, Upper Saddle River, NJ. Chapter 6. A compact treatment at an intermediate level.

the Chemical Bond." He considered the following constraints on reasonable protein fiber structures:

- bond lengths and angles are as found in X-ray diffraction studies of amino acids and small peptides;
- free rotation is possible *only* about the bond from the alpha carbon to the amino group (NH) and the bond from the alpha carbon to the carboxyl group (C=O) (see Fig. 3.1);
- some interaction between groups must exist to stabilize any structure. Hydrogen bonds—noncovalent bonds between −NH...O=C—seemed the most likely candidates.

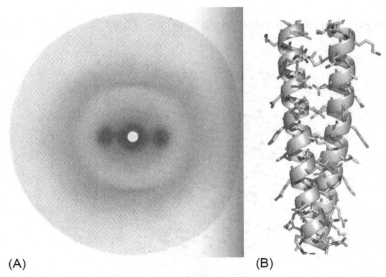

(A)   (B)

FIG. 4.1 X-ray diffraction studies of keratin. (A) The fibers used for the X-ray diffraction pattern shown came from a lock of hair of the composer Wolfgang Amadeus Mozart. The X-ray diffraction photograph was taken by Dr. Elwyn Beighton. The fiber diagram shows rather indistinct arcs rather than sharp spots. This is partly because of the irregular order of the amino acids in keratin, but also because the protein chains are coiled up in a complicated way, as shown diagrammatically in part (B). Each chain is wound into a tight helix, and then pairs of chains are wound round each other – the picture here shows just a short length of such a "coiled-coil". *(Adapted from http://www.leeds.ac.uk/heritage/Astbury/Mozarts_Hair/ index.html).*

With these constraints, using folded cardboard models, Pauling deduced that only a few regular chain structures were feasible. Two of these, shown in Figs. 4.2A and 4.2B, were termed the **alpha** (α) **helix** and the **beta** (β) **sheet**, respectively. The first predicted an X-ray diffraction pattern consistent with Astbury's alpha keratin, the second matched with beta keratin. The discovery of these structures was a triumph of pure thought. Each satisfies the criteria listed earlier, but in quite different ways. The α-helix has 3.6 amino acid residues per turn of the helix. Stability is provided by hydrogen bonds between the N−H in the backbone and the COO− 13 residues farther along. The β-sheet structure, on the other hand, is made by lying extended chains side by side, with hydrogen bonds *between* the chains. Actually, there are two possible β-sheet structures, depending on whether adjacent chains run in the same (parallel) or opposite (antiparallel) directions (see Fig. 4.2B for details of these structures). It should be noted that a few other regular structures are theoretically possible, and sometimes observed, but are apparently less stable, and are rare.

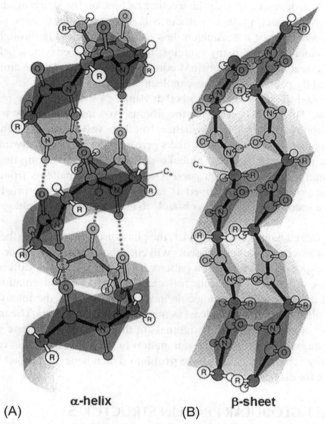

α-helix
(A)

β-sheet
(B)

FIG. 4.2 The α-helix and the β-sheet: the two most common regular secondary structures in proteins. They differ mainly in how the hydrogen bonds stabilizing the structures are formed. (A) In the α-helix the H bonds are within a single length of coiled polypeptide chain and are oriented almost parallel to the helix axis. The number of residues in this helix is 3.6 per turn of the helix which corresponds to 0.54 nm per turn. (B) In the β-sheet, the hydrogen bonding occurs between spatially adjacent chains. These could belong to the same polypeptide chain folded back upon itself or to two individual chains of proteins that have multi-subunit structures (see quaternary structure). In the β-sheet, the hydrogen bonds are almost perpendicular with respect to the chain direction. *(From Molecular Biology: Structure and Dynamics of Genomes and Proteomes by Jordanka Zlatanova and Kensal E. van Holde. Reproduced by permission of Taylor and Francis Group, LLC, a division of Informa plc. Copyright 2015.)*

GLOBULES

Contemporary with Astbury's work on fibrous proteins, evidence was accumulating that more compact structures were the norm for the vast majority of protein types. A major part of this evidence came from the sedimentation studies of Svedberg and a few other laboratories that had copies of his ultracentrifuges. As described in section "Sedimentation" in Box 3.1 the velocity of sedimentation depends, in addition to other factors, on the shape of the macro-molecule. A compact, globular particle will sediment faster than an elongated one of the same mass; it encounters less resistance in moving through the so-lution. It was found that many proteins, including most enzymes, behaved as nearly spherical objects. Dimensions commonly indicated that the amino acids must be tightly packed within these molecules.

This raised the question as to whether Pauling's structures were relevant to all proteins. They clearly were for the fibrous proteins, where they were con-firmed by the X-ray diffraction evidence. But there was no such evidence in the case of globular proteins. Whereas fibers give very simple and easily interpreted diffraction patterns, those from globular proteins were, for a long time, found to be uninterpretable. Although good X-ray diffraction patterns from crystals of globular proteins were observed as early as 1934, it was soon realized that there was a seemingly intractable obstacle to interpretation—the phase problem (Box 4.1).

As explained a bit more in Box 4.1, the phase problem relates to the fact that different atoms in a complex molecule will contribute differently to the intensity of a given spot on the diffraction pattern. They will contribute differently be-cause of the different **phase** of the reflected rays. The total information needed to deduce the structure of a molecule in a crystal lattice is the intensities *and* phases of all reflections—or at least as many as can be observed. The intensities are easy—just measured by the brightness of the spots. But the phase summary at each spot, which results from summation of waves at that spot, cannot be immediately seen. This is the **phase problem** which bedeviled protein crystal-lographers for decades.

THE FIRST GLOBULAR PROTEIN STRUCTURES

The final solution was due primarily to the patient, long-unrewarded efforts of a quiet Austrian scientist named Max Perutz. He had an eventful life. Born in Vienna in 1914, he had come to Cambridge England in 1936 to work with the famous crystallographer John Desmond Bernal. When Germany occupied Austria in 1938, Perutz elected to stay in England, and was able to bring his family there. Unfortunately, when the war started, Perutz was immediately clas-sified as an "enemy alien" and interned. After some time, Bernal secured his release to work on Project Habbakuk, one of the strangest and least known pipe-dreams of World War II. It was proposed to build a giant "ice island" floating in the mid-Atlantic to serve as a landing field for warplanes being delivered to Britain. Needless to say, nothing came of Habbakuk, but at least it kept Perutz out of prison.

After the war, Perutz returned to Cambridge, determined to solve the structure of the important blood protein, hemoglobin (recall that hemoglobin was the first protein to be crystallized). His assistant, a young British scientist named John Kendrew, chose to work on **myoglobin**, a smaller relative of hemoglobin that stores oxygen transported by hemoglobin. In fact, hemoglobin is constructed of four chains very like myoglobin. But the first aim of all this work was to devise a way to solve the "phase problem." That occupied nearly 10 years of constant trials and labor. Finally, in the mid-1950s, a method emerged: if an atom in the molecule could be substituted by a heavy atom, in the exact same place in every molecule in the crystal, all reflections involving that atom would be altered. This could then be used—especially if several atoms were replaced—to deduce the phases of other reflections and hence solve the molecular structure of the crystallized protein.

It was myoglobin, the smaller protein, which was solved first. We must qualify the word "solved," for protein structures can be visualized at different levels of detail (often referred to as "levels of resolution"). Resolution depends on how many spots in the diffraction pattern are taken into account (Box 4.1). The first structure of myoglobin, published in 1958, used 400 reflections and gave a resolution of only 0.6 nm. It could show only regions of high electron density, some of which might or might not be alpha helices. But even this limited information was useful. Striking was the complete lack of symmetry. As Jim Watson said, "it looked like a lady's hat." Over the next few years the resolution was greatly improved; by 1961, 0.15 nm resolution was attained, using 20,000 reflections (Fig. 4.3). This was sufficient to resolve details of side-chain groups; the protein sequence could be read from the structure! Myoglobin (and soon thereafter hemoglobin) turned out to be rich in α-helices, with dimensions just as Pauling had predicted.

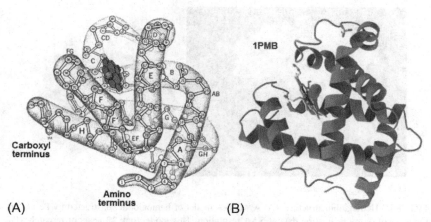

FIG. 4.3 Sperm whale myoglobin. (A) Notation of amino-acid residues in helical and non-helical regions of the protein. (B) High-resolution structure constructed from atomic coordinates deposited in the Protein Data Bank under ID 1PMB. Ribbons represent secondary structures that are twisted in α-helices. *(Part A from Perutz, M.F., Muirhead, H., Cox, J.M., Goaman, L.C., 1968. Three-dimensional Fourier synthesis of horse oxyhaemoglobin at 2.8 Å resolution: the atomic model. Nature 219, 131–139, Fig. 1, with permission from Nature Publishing.)*

A schematic view of the hemoglobin molecule is shown in Fig. 4.4. We can describe this structure at four levels of organization. First, there are four polypeptide chains of two sequence variants, called alpha and beta (not to be confused by Pauling's alpha and beta!). These constitute the **primary structure**. Each chain contains a good deal of α-helix; this local, regular folding is termed **secondary structure**. The α-helices are folded back and forth to produce the **tertiary structure** of each subunit. Then the four subunits (two of each variant) associate together by noncovalent interactions to form a roughly tetrahedral **quaternary structure**. Not all proteins form multisubunit quaternary structure, but noncovalent interactions between different proteins are undoubtedly common in the environment of the cell. We are beginning to recognize the importance of these and will discuss a number in later chapters.

Over subsequent years, thousands of protein structures have been elucidated by X-ray diffraction. They turn out to exhibit an astounding variation in structure. Some are mostly alpha helical, some mostly beta sheet, many containing both kinds of structures, and some have little or no organized structure. These latter disorganized proteins are capable of taking on a variety of conformations, and thereby interacting with a wide range of other molecules; they are sometimes called "promiscuous." As the number of solved protein structures has accumulated, an interesting general principle has emerged: many proteins are constructed of two or more **domains**—independently folded regions of a peptide chain. Certain domains may be recognized in several proteins, where they

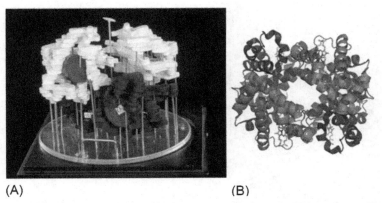

(A)   (B)

FIG. 4.4   Hemoglobin structure. (A) A wooden model of hemoglobin constructed by Perutz on the basis of X-ray diffraction data at 5.5Å resolution. This model took 22 years of research. (B) Structure constructed from recent atomic coordinates deposited in the Protein Data Bank under ID 1GZX. The proteins α and β subunits are in two different shades of *gray* and the iron-containing heme groups in *gray lines*. *(Part A from Perutz, M.F., Muirhead, H., Cox, J.M., Goaman, L.C., 1968. Three-dimensional Fourier synthesis of horse oxyhaemoglobin at 2.8 Å resolution: the atomic model. Nature 219, 131–139, Fig. 3, with permission from Nature Publishing.)*

often perform similar functions. It seems as if proteins can have "interchangeable parts," in order to combine capabilities.

What determines the tertiary structure of a protein? What holds the polypeptide chain in so specific a conformation? This question is intimately connected to the phenomena of protein **denaturation** and **renaturation**. As early as 1900, it was recognized that many proteins would precipitate, if heated too high or treated with acids or bases. In many cases the protein could not be resolubilized and was said to be "denatured" (consider a boiled egg). The first clear understanding of denaturation and its implications for protein stability came from a seemingly unlikely source—China in 1931. In a series of papers published in the Chinese Journal of Physiology, Hsien Wu presented a remarkably advanced view of globular protein structure. He visualized it as a highly organized arrangement of the polypeptide chain, stabilized by noncovalent interactions between amino acids. Denaturation, Wu proposed, involved the breaking of these interactions and the unraveling of the compact structure. This is essentially what we believe today. Wu's work was unknown in the west; in 1936 Linus Pauling and Alfred Ezra Mirsky published essentially the same model, wholly unacquainted with Wu's precedence.

For some time, there was the question as to whether the structure of a protein in a crystal, where it is crowded cheek-by-jowl with neighbors was the same as its conformation when in solution. This has been answered in general by the recent development of **nuclear magnetic resonance** (**NMR**) methods that can determine protein structure in solution nearly as accurately as X-ray diffraction. In most cases the two techniques yield superimposable results.

There are about 20,000 different proteins coded for by the human genome. Elucidation of that complexity and the enormously greater potential complexity of their interactions has become the topic of a new subdiscipline of biochemistry—**proteomics**. This, in turn, is closely linked to **genomics**, the analysis of the total genetic content of organisms. To understand this, we must step back and see how, during the rise of biochemistry, genetics was undergoing its own origins and evolution.

EPILOGUE

We have followed the progress of biochemistry—as traced by the more and more comprehensive analysis of proteins—to the early 1960s. By that point the understanding of protein sequence and three-dimensional structure had become deep enough to allow the formulation of a wholly new and fascinating question: how was the vast amount of information necessary to specify sequence and structure of thousands of different proteins passed from cell to cell or generation to generation? But this is the topic of **genetics**, which had hitherto been a wholly separate science. So now we must change direction for a while and explore the background and rise of genetic research.

FURTHER READING

Books and Reviews

Anfinsen, C.B., Scheraga, H.A., 1975. Experimental and theoretical aspects of protein folding. Adv. Protein Chem. 29, 205–300. Concerned with spontaneous folding.

Branden, C., Tooze, J., 1999. Introduction to Protein Structure, second ed. Garland Publishing, New York. An excellent treatment, a bit dated now.

Dunker, A.K., Oldfield, C.J., Meng, J., Romero, P., Yang, J.Y., Chen, J.W., Vacic, V., Obradovic, Z., Uversky, V.N., 2008. The unfoldomics decade: an update on intrinsically disordered proteins. BMC Genom. 9 (Suppl. 2), S1. From pioneers in this area.

Orengo, C.A., Thornton, J.M., 2005. Protein families and their evolution – a structural perspective. Annu. Rev. Biochem. 74, 867–900. Good overview.

Pauling, L., 1960. The Nature of the Chemical Bond, third ed. Cornell University Press, Ithaca, NY. Pauling's classic work.

Pestko, G.A., Ringe, D., 2003. Protein Structure and Function (Primers in Biology). New Science Press, Waltham, MA. Introduction to the general principles of protein structure, folding, and function. Also, provides the conceptual basis of inferring structure and function from genomic sequence.

Tanford, C., Reynolds, J., 2001. Nature's Robots: A History of Proteins. Oxford University Press, Oxford, UK. A very comprehensive and readable history of protein chemistry; probably the best source to date.

Classic Research Papers

Pauling, L., Corey, R.B., Branson, H.R., 1951. The structure of proteins; two hydrogen-bonded helical configurations of the polypeptide chain. Proc. Natl. Acad. Sci. USA 37, 205–211. A theoretical derivation of the two helical configurations of polypeptide chains based on the postulate that all amino acids in a chain are equivalent (except for the differences in the side chain R).

Perutz, M.F., Muirhead, H., Cox, J.M., Goaman, L.C., 1968. Three-dimensional Fourier synthesis of horse oxyhaemoglobin at 2.8 Å resolution: the atomic model. Nature 219, 131–139. The first three-dimensional model of hemoglobin, focusing on features that are particularly significant for its function, such as the surrounding of the heme groups and the contacts between the four protein subunits.

Yu, H., Braun, P., Yildirim, M.A., Lemmens, I., Venkatesan, K., Sahalie, J., Hirozane-Kishikawa, T., Gebreab, F., Li, N., Simonis, N., Hao, T., Rual, J.F., Dricot, A., Vazquez, A., Murray, R.R., Simon, C., Tardivo, L., Tam, S., Svrzikapa, N., Fan, C., de Smet, A.S., Motyl, A., Hudson, M.E., Park, J., Xin, X., Cusick, M.E., Moore, T., Boone, C., Snyder, M., Roth, F.P., Barabasi, A.L., Tavernier, J., Hill, D.E., Vidal, M., 2008. High-quality binary protein interaction map of the yeast interactome network. Science 322, 104–110. An interaction map between pairs of proteins is created using high through-put methods. The map covers ~20% of all yeast binary interactions. There is a highly significant clustering between essential yeast proteins.

Wu, H., 1995. Studies on denaturation of proteins. Adv. Protein Chem. 46, 6–26. This is a reprinting of Wu's 1931 paper on protein denaturation in The Chinese Journal of Physiology. A prescient work.

Chapter 5

The Origins of Genetics

PROLOGUE

In the previous four chapters, we have briefly described the origins and development of Biochemistry as one of the two major branches in life sciences that contributed to the rise of molecular biology. We have tracked this development by following the growing understanding of the chemical and physical structure of proteins, the basic machinery used by cells to perform numerous functions. There is, however, much more to molecular biology than structure and biochemistry. A true molecular biology must also explain how the information needed to produce biological structures and processes is stored, expressed, and transmitted from one generation to the next. This task lies in the province of genetics. Much of genetics (what we now call "classical" genetics) was developed before anything was known of the molecular processes involved, yet it provided a vital and important impetus and direction for the new science of molecular genetics. Thus, it is appropriate to provide a brief chapter on this background.

CLASSICAL GENETICS AND THE RULES OF TRAIT INHERITANCE

Genetics is the study of inheritance. From earliest times, philosophers and plant and animal breeders have recognized that there are similarities between parents and their offspring, but any quantitative explanation for how traits are transmitted and which are favored was long in coming. Existing theories postulated such ill-defined ideas as the "mixing of genetic fluids" to explain the traits of offspring. It was not until the middle of the 19th century that the first rigorous analysis was carried out. Remarkably, this was not done by any of the eminent biologists of the day, but by an obscure Austrian monk, the Augustinian Friar Gregor Mendel.

FRIAR GREGOR MENDEL PLANTS SOME PEAS

Mendel was born (as Johann Mendel) in 1822. At the age of 21 he entered the Augustinian monastery in Brünn (now Brno), Austro-Hungarian Empire (now in the Czech Republic) and took the name Gregor. He must have shown intellectual abilities early, for the order sponsored his four semesters of study in physics in Vienna.

The Evolution of Molecular Biology. https://doi.org/10.1016/B978-0-12-812917-3.00005-X

Mendel tended gardens at the Monastery at Brno and tried to understand how visible phonotypic features (traits) of the small garden pea were transmitted from one plant generation to the next. The pea plant was an excellent choice because peas can be raised rapidly and exhibit clearly recognizable traits that can breed true for many generations (i.e., many generations continue to exhibit the same trait). In addition, peas are capable of either self-fertilization or cross-fertilization (Fig. 5.1). It is believed that Mendel raised and examined (over eight years) more than 28,000 pea plants. The unusual (for this field at the time) aspect of Mendel's work was his careful quantitation of the outcome of every breeding experiment. He painstakingly kept records of all his observations for nearly a decade and performed statistical analysis of the results.

In his experiments, Mendel would first choose parental stocks of plants exhibiting contrasting traits in a particular character (yellow vs green seeds, for example; these morphologically observable characters are referred to as **phenotypes**). These plants he selected after propagating the plants for several generations to ensure that all the progeny exhibited the same phenotype, in other word the parental stocks were "breeding true" in self-fertilization. These we will refer to as the parental phenotypes, the P_1 generation (Fig. 5.2). When these were cross-fertilized, it was always observed that only one of the two alternate traits was expressed in the progeny, called the F_1 (Filial 1) generation. In our example, the trait expressed in F_1 is yellow (Fig. 5.2). This showed Mendel that one trait

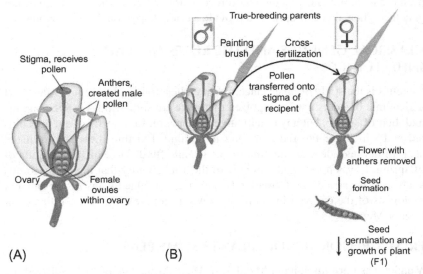

(A)   (B)

FIG. 5.1   Mendel's experiments with garden peas. (A) Anatomy of the pea flower: peas are self-fertilizing but can also be cross-fertilized, as illustrated in (B). Mendel followed the inheritance pattern of several contrasting (antagonistic) forms of particular traits: seed color and shape, flower color, pod color and shape, stem length, and flower position. *(From Molecular Biology: Structure and Dynamics of Genomes and Proteomes by Jordanka Zlatanova and Kensal E. van Holde. Reproduced by permission of Taylor and Francis Group, LLC, a division of Informa plc. Copyright 2015.)*

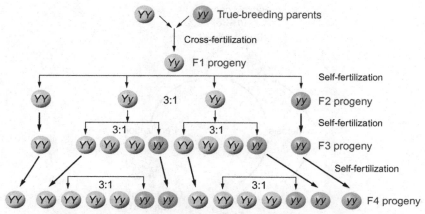

FIG. 5.2   Distribution of a pair of contrasting traits (seed color) over four generations. Capital *Y* stands for yellow color, lowercase *y* stand for green color. For each trait, the plant carries two copies of a unit of inheritance (two alleles of a gene in contemporary understanding). The trait that appears in all F_1 hybrids is dominant, the antagonistic trait that remains hidden in F_1 but reappears in F_2 is recessive. In this specific example, yellow color is dominant, green is recessive. The schematic also illustrates the difference between homozygous and heterozygous individuals. The homozygous individuals of each generation breed true (thick arrows), whereas heterozygous do not (thin arrows).

(yellow seed) was **dominant** over the other (green seed). In the next series of experiments members of the F_1 generation were self-hybridized. Now the other trait, the **recessive** trait, reappeared, but in only ¼ of the F_2 progeny. This meant that there must be **units of heredity** that could be distributed according to fixed rules and dictated the dominant and recessive states. In Fig. 5.2, the determinant of the dominant trait is designated Y, the recessive y. We now call these units of heredity **genes**, but the term was not used until decades after Mendel's time.

In order to understand in modern terms what Mendel was observing, we must jump ahead and take a detour into **sexual reproduction** and the two types of cell divisions that take place during the reproduction cycle (Fig. 5.3). Most cells of the plant or animal body (somatic cells, from the Greek *soma*, body) are **diploid**, meaning that they carry two homologous chromosomes. These diploid cells are reproduced through a cell division process called **mitosis**. In mitosis, the genetic material (DNA) is first duplicated, then compacted into condensed chromosome structures, each of which contains only one long DNA molecule. Finally, the chromosomes are separated equally into the two daughter cells, which are now exact diploid copies of the mother cell. The gametes, on the other hand, are **haploid**, carrying only one chromosome of a particular type, and thus one set of genes.

Gametes (sperm or ova) arise through a special division process called **meiosis**. In meiosis, the initial duplication of the genetic material and its compaction into chromosomes is followed by two successive divisions of the chromosomes, in a way that produces four (rather than just two, as in mitosis) cells.

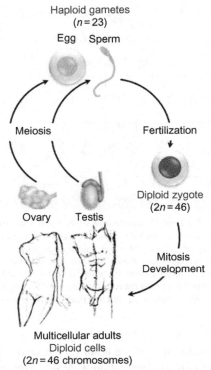

Haploid gametes
($n = 23$)

Egg   Sperm

Meiosis

Fertilization

Ovary   Testis

Diploid zygote
($2n = 46$)

Mitosis
Development

Multicellular adults
Diploid cells
($2n = 46$ chromosomes)

FIG. 5.3   **A conceptualized schematic of the sexual reproduction cycle in higher eukaryotes.** Note the transitions between cells containing the diploid number of chromosomes (46 in humans) and those with haploid number of chromosomes (23 in humans). The first transition occurs during the formation of the female and male gametes (egg and sperm, respectively) and involves a reduction division known as meiosis. The haploid gametes then reunite during fertilization to form the diploid zygote (fertilized egg). All cells of an adult organism are formed from further divisions known as mitosis.

Each gamete contains only one set of chromosomes (rather than the two in the somatic cells). As a result of the combination of sperm and ovum in sexual reproduction, the progeny contain two copies of each gene.

MENDEL FORMULATES THE TWO LAWS OF INHERITANCE

From these experiments, Mendel formulated the two "laws" (really hypotheses) that constitute the basis of classical genetics. **Mendel's first law** (the **law of segregation**) states that the two alleles for each trait separate (segregate) during gamete formation, then unite at random, one from each parent, at fertilization. To facilitate understanding of the first law in modern terms, we break it into several statements.

- Variation in phenotype (observable characteristics or traits) is explained by the existence of alternate versions of genes. These versions are called **alleles**.

- The alleles of each gene segregate, independently, one to each gamete.
- Every individual inherits two alleles of each gene, receiving one gamete from each parent.
- If the alleles differ, one will be dominant, one recessive. If the individual is heterozygous in an allele (carries both versions), only the dominant allele and trait will be expressed in the first generation.

But what happens if one cross-breeds peas that differ in two traits? Does the segregation of alleles for one trait affect the other? Mendel also carried out such experiments and derived what is called **Mendel's second law** (the **law of independent assortment**). The law states that during gamete formation, the segregation of the alleles of one allelic pair is independent of the segregation of the alleles of another allelic pair. In other words, traits segregate independently; there is no linkage between genes for different traits. This happens not to be always true, as we shall see.

MENDEL'S LAWS HAVE EXTENSIONS AND EXCEPTIONS

As with many great scientific breakthroughs, the true situation has proved to be more complicated that the initial insight suggested. It took years to recognize and properly understand the many examples where simple Mendelian genetics is excepted. In general, the extensions to Mendel's laws can be classified in two groups, depending on whether a trait is encoded by a single gene or by many genes (multifactorial inheritance).

In the **single-gene inheritance group**, there are three major extensions. First, dominance is not always complete. We now distinguish between **incomplete dominance**, where the hybrid resembles neither parent, and **codominance**, where neither allele is dominant, with the F1 hybrid showing traits from both true-breeding parents. Second, a gene may have more than two alleles. There are numerous examples of such multiallele genes, with the most extreme case being the olfactory genes, which have ~1300 alleles, only one of which is expressed. Usually cells express both alleles of a gene. Finally, one gene may contribute to several visible characteristics; this phenomenon is known as **pleiotropy** (from the Greek *pleion*, meaning "more," and *tropi*, meaning "to turn, to convert"). A classic example of pleiotropy is found in sterile males from among the aboriginal Maiori people of New Zealand. These men are sterile and have respiratory problems. The gene's normal dominant allele specifies a protein needed in both cilia and flagella; in men who are homozygous for the recessive allele, cilia and flagella do not function properly, affecting their abilities to both clear mucus from their respiratory tract and produce motile sperm.

In the **multifactorial inheritance group**, two or more genes can interact to determine a single trait. Thus, novel phenotypes can emerge from the combined action of the alleles of two genes. Also, there are continuous (quantitative) traits that vary continuously over a range of values. These traits are polygenic and show the additive effects of a large number of genes and their alleles. A good example is height and skin color in humans.

Finally, it is now clear that the environment can affect the phenotypic expression of a genotype. When environmental agents cause a change in phenotype that mimics the effects of a mutation in a gene, we talk about **phenocopying**. A painful example of this phenomenon is the effect of the sedative drug thalidomide. If taken by pregnant women, this drug produced a phenocopy of a rare dominant trait called phocomelia, which disrupts limb development in the fetus.

Perhaps the most significant exception is the fact that Mendel's second law is not generally correct. There are many cases in which genes are found not to segregate independently. Such a pair of genes are said to be **linked**; the physical basis for gene linkage is close proximity on chromosomes. It seems likely that Mendel did not observe linked genes because many of the pairs of genes involved in his studies lay far apart, even on different chromosomes. It is also necessary to note that some critics have claimed that Mendel could have occasionally "tweaked" the data a bit. Regardless of the truth or falsity of this, Mendel essentially founded genetics.

MENDEL WAS LONG IGNORED

Although Mendel published a detailed account of his experiments and their interpretation in 1867, it was almost wholly ignored for over 30 years. In that period, the work received only three citations. To some extent, this was Mendel's fault—he chose to publish in a truly obscure local Journal, the Verhandlungen des Naturforschenden Vereins Brünn. He did send reprints to illustrious biologists (including Charles Darwin), but these seem to have been ignored. Perhaps it is because mathematical analysis of biological problems was almost unknown at the time.

It was not until the first decade of the 20th century that Mendel was "rediscovered." Then a number of scientists almost simultaneously came upon the long-neglected paper. The British biologist William Bateson called it to the attention of the Royal Society, and an English translation appeared in the Proceedings of the Royal Horticultural Society in 1901. One can truly consider that genetics, as a science, only began at about this date; curiously, it is virtually synchronous with the true birth of protein chemistry (Chapter 3).

DARWIN, MENDELISM, AND MUTATIONS

Charles Darwin, undoubtedly the greatest of all 19th century biologists, plays a surprisingly small role in the evolution of molecular biology. This is because Darwin's methods were primarily observational, and his view of evolution avoided speculation as to mechanisms. An important contribution is an indirect one; an early critique of Darwinism lay in the concern that there was simply not enough geological time, as assumed then, to allow for the whole course of evolution. Darwin postulated a slow natural variation in species augmented by natural selection. A remarkable insight into the problem came from an early proponent of Mendel, the Dutch scientist Hugo de Vries. In 1901 de Vries published "Species and Varieties; Their Origin by Mutation." This was the first introduction of the concept of **mutations**, sudden changes in the genetic material that could be then propagated by Mendelian inheritance, and selected for or against. This radical idea provided the impetus and direction for all of the genetic research to immediately follow.

GENES ARE ARRANGED LINEARLY ON CHROMOSOMES AND CAN BE MAPPED

Although a number of scientists were involved, the rapid development of genetics in the early 20th century can be ascribed to a great degree to the work of the American, Thomas Hunt Morgan. Like Mendel, Morgan was distinguished by an ability to see the important general laws in a mass of complex data.

Morgan spent most of his career at Columbia University. Rather than working alone, as did Mendel, Morgan gathered about him an unusually talented group of students and postdoctoral researchers, many of whom also went on to make major contributions to genetics. These included Alfred Sturtevant, who constructed the first genetic map of a chromosome (see below); George Beadle, who with Edward Tatum first related genes to proteins; Theodosius Dobzhansky, who showed how mutations could drive evolution; and Hermann Muller, who discovered the mutational effects of short-wavelength radiation. The list could be extended for further academic "generations." It should be noted that some historians credit Morgan with revolutionizing the way in which research groups operate. Up to Morgan's time, the European model, with a rigid, hierarchical structure dominated by an almost inaccessible Professor was emulated throughout the scientific world. In stark contrast, Morgan's laboratory had an informal, relaxed ambience; visiting scientists were often shocked to find all participants on a first-name basis. Gradually, this became the norm in American universities, and to some extent, elsewhere.

In the early part of the 20th century, Morgan began working with fruit flies, *Drosophila melanogaster*. This was an even better choice than Mendel's peas because the flies breed very rapidly and could be raised in large numbers in very little space. This was important because Morgan was dependent, for much of his work, on the occurrence of rare, spontaneous mutations. These were the source of the changes in the genes that produced modified alleles and hence different phenotypes. Thus, the discovery, by Herman Joseph Muller in Morgan's lab, that mutations could be induced by X-rays or other damaging radiation, was very helpful. Muller was awarded the 1946 Nobel Prize in Physiology or Medicine "for the discovery of the production of mutations by means of X-ray irradiation."

Contrary to Mendel's observations, Morgan found that a number of traits of the flies appeared to be genetically linked. The difference with Mendel's conclusion may be explained by the fact that many of the pea traits studied by Mendel corresponded to genes on different chromosomes, which would be expected to segregate independently, whereas Morgan was initially concentrating on genes on the same (female X) chromosome. In any event, Morgan observed linked transmission of a number of genes. He hypothesized that the lack of linkage in some cases must result from recombination of alleles (Fig. 5.4A).

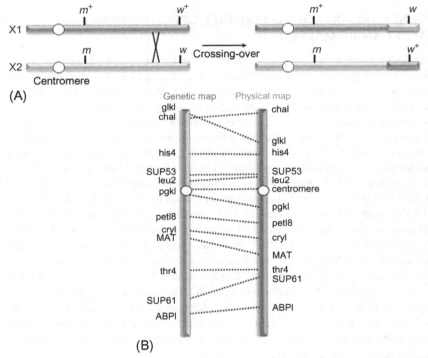

FIG. 5.4 Genetic recombination and genetic maps. (A) Recombination between the two X chromosomes of the female *Drosophila* fly. Chromosome X1 carries two wild-type alleles: m^+ codes for normal wings and w^+ determines red eyes. Chromosome X2 carries two mutant alleles, coding for miniature (m) wings, and white (w) eyes. During egg formation, a crossing-over (recombination) event occurs somewhere between these two genes on the two chromosomes, resulting in two recombinant chromosomes, each of which carries a mixture of the parental alleles. This process creates a new combination of alleles, hence the name recombination. (B) Genetic and physical maps of *Saccharomyces cerevisiae* chromosome III. The genetic map was constructed by determining the frequency of recombination in genetic crosses; the physical map was determined by DNA sequencing. Despite some discrepancies between the two maps, their overall similarity is impressive. Note that the order of the upper two markers (identifiable genes) has been incorrectly assigned on the genetic map; the relative positions of some markers are also somewhat different on the two maps. *(Part B adapted from Oliver, S. G., et al., 1992. Nature 357, 38–46, Fig. 3, with permission from Macmillan Publishers, Ltd.)*

Furthermore, he noted that the probability of such recombination must increase with the distance between the two genes in the chromosome. Thus, the degree of linkage must measure gene separation. Then, something remarkable happened. Alfred Sturtevant, at that time a student working in Morgan's laboratory, realized that this fact allowed "mapping" of genes on chromosomes. He skipped his assigned homework one night to produce the first genetic map (1913). Soon this was extended to many *Drosophila* genes, and a new paradigm emerged: genes are arranged linearly on chromosomes. An example genetic map is presented in Fig. 5.4B. In 1933, Morgan was awarded the Nobel Prize in Physiology or Medicine "for his discoveries concerning the role played by the chromosome in heredity."

WHAT DO GENES DO, AND WHAT ARE THEY MADE OF?

Although many classical geneticists seemed content to regard genes as abstract entities that obeyed certain mathematical laws, others sought for the substance and its mode of function. What genes might do was suggested as early as 1909, by the British scientist Sir Archibald Garrod. Garrod was studying an alarming but relatively harmless human anomaly called alkaptonuria, in which urine rapidly turns dark colored on contact with air. The surprising result of studies of families exhibiting this condition was that it was inherited in strictly Mendelian fashion. Garrod noted the same pattern in a number of conditions, including albinism. He correctly guessed that lack or malfunction of a specific enzyme, related to a specific gene, was the explanation. Like Mendel, Garrod had to wait for recognition: his work was neglected for nearly 30 years, until George Beadle and William Tatum carried out similar analyses using the mold, *Neurospora*. They stated the dictum "One gene, one enzyme," which henceforth defined the problem.

The connection between genetics and protein sequence was strengthened by some remarkable results on **sickle-cell anemia**. This debilitating blood condition is characterized by red blood cells taking on a "sickle" shape (see Fig. 15.3), which makes them fragile and prone to block capillaries. In 1949, James Neel and E. A. Beet working independently showed that the sickle-cell trait was inherited in a classic Mendelian manner. In the same year, Linus Pauling found that the hemoglobin of patients homozygous for the trait differed in electrophoretic mobility from that of normal individuals. Furthermore, those that were heterozygous showed both bands (see Fig. 15.3). In later years, it was demonstrated that a single amino acid substitutions in one of the hemoglobin chains conferred a tendency for hemoglobin molecules to form long rods, which distorted the red cells into the sickle shape.

Despite the rapid advances in classical genetics in the early years of the 20th century, the chemical nature of genes and their mode of function remained obscure. Genes gained some physical reality through the examination of polytene chromosomes. These are parallel aggregates of a number of

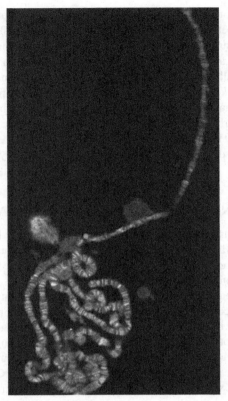

FIG. 5.5 Polytene chromosomes with the typical banding patterns. The banding patterns of *Drosophila* polytene chromosomes were depicted by Calvin Bridges as early as 1935 and are still widely used to identify chromosomal rearrangements and deletions. Fluorescent image of the *Drosophila* salivary gland chromosomes stained for DNA and two proteins (each by a different shade). *(From Wikipedia, LPLT/Wikimedia Commons.)*

identical chromosomes, found in the salivary glands of some insects, including *Drosophila* (Fig. 5.5). Upon proper staining, they show a banded pattern under the light microscope; some bands can be correlated with genes mapped by the Sturtevant-Morgan technique. Such studies reinforced the image of genes as physical objects—but of what were they made?

Search for the answer got off to a bad start. This same period saw the beginnings of protein studies, and the function of proteins as enzymes was well established. Probably because proteins were the most complex molecules known to that date, and could do so many things, most researchers before about 1940 assumed that genes were proteinaceous in composition. It was also becoming evident, however, that genes must dictate protein structure. After the brilliant work in several laboratories on sickle-cell anemia, a genetic disease that affects millions of people around the world, this dictum had to be modified to "one gene-one polypeptide chain." Today we realize that not only protein sequences but also nonprotein-coding RNA sequences are dictated by genes, leading to further attempts to properly define a gene. As we shall see in Chapter 14, the definition of the term "gene" is still evolving, with new definitions proposed as a result of studies on the human genome.

EPILOGUE

In the brief period from 1900 until about 1920, genetics grew from a neglected seedling into a robust science. It still lacked mechanistic underpinnings; even the nature of the postulated genes was wholly unknown. The first guess, that genes were proteins, would soon prove wrong. Instead, scientists had to turn to another whole class of biological macromolecules, the **nucleic acids (poly-nucleotides)**. Before discussing this great jump in concept, we must see what these substances are, and what unique properties they possess.

FURTHER READING

Books and Reviews

Allen, G.E., 1978. Thomas Hunt Morgan: The Man and His Science. Princeton University Press, Princeton, NJ. Complete biography of Morgan based on few available sources, including his extensive correspondence with other scientists.

Henig, R.M., 2000. The Monk in the Garden: The Lost and Found Genius of Gregor Mendel, the Father of Genetics. Houghton Mifflin, Boston, MA. A vivid story about the life of Mendel, described by some critics as a 'provocative portrait' of the monk and his work.

Morgan, T.H., Sturtevant, A.H., Muller, H.J., Bridges, C.B., 1915. The Mechanisms of Mendelian Heredity. Henry Holt and Company, New York, NY. A summary of the major findings.

Otto, S.P., 2008. Sexual reproduction and the evolution of sex. Nature Educ. 1, 182. Treatise of sexual reproduction from an evolutionary point of view.

Russell, P.J., 2009. iGenetics: A Molecular Approach, third ed. Benjamin Cummings, San Francisco, CA. Excellent textbook that reflects the increasing molecular emphasis in today's experimental study of genes; provides an appreciation for classic experiments and helps students to develop problem-solving skills.

Sturtevant, A., 2001. A History of Genetics. Cold Springs Harbor Laboratory Press, Cold Springs Harbor, NY. A highly-praised book on the history of genetics.

Classic Research Papers

Mendel, J.G., 1865. Versuche über Pflanzenhybriden (Experiments in plant hybridization). . The original work of Mendel, presented at Brünn natural history society (Brno, Presently in the Czech Republic). First translated into English by William Bateson in 1901. A translation, with commentary, is to be found in James, A., Peters, E. D., 1959. Classic Papers in Genetics. Prentice Hall, Englewood Cliffs, NJ.

Morgan, T.H., 1910. Sex limited inheritance in Drosophila. Science 32, 120–122. Describes the results of cross-breeding of wild-type red-eyed fruit flies with white-eyed mutants. The new phenotype appeared only in male progeny, that is, its transmission was sex-limited.

Sturtevant, A.H., 1913. The linear arrangement of six sex-linked factors in *Drosophila*, as shown by their mode of association. J. Exp. Zool. 14, 43–59. Describes the derivation of the first genetic map, with all of its genes in their correct position; it also clearly lays out the logic for genetic mapping using experimentally determined frequencies of cross-overs between genes on chromosomes.

Chapter 6

Nucleic Acids

PROLOGUE

As the physical reality of genes gained credence through their mapping, the mystery of their substance deepened. For example, if proteinaceous genes dictated protein structure, what dictated these genes? Another gene made of protein? An infinite regression would seem necessary. Apparently, some other substance must constitute the genetic material—but what? At this point scientists began reconsidering a very old idea that mysterious components of the cell nucleus, called **nucleic acids** might be the sought-for substances. To describe this search, we must first tell a bit about these materials and their history.

MIESCHER'S MYSTERIES

In Chapter 2 we briefly described Johannes Miescher's pioneering investigations of the cell nucleus. The substance he recovered had an unusually high phosphorus content—about 2.5%. Since proteins contain very little phosphorus—usually 1% or less—this was clearly something very different. Indeed, it was so unusual that Felix Hoppe-Seyler, Miescher's research director, insisted on repeating the experiments himself before he would allow publication. Miescher then moved to Basel, Switzerland, and used salmon sperm as a source. A combination of acid and alkali treatments produced two fractions: one was a basic, protein-like substance, and the other, acidic and rich in phosphorus. The latter was termed "Nucleïn-säure" (nucleic acid) and must have been close to what we recognize as DNA, for it had a phosphorus content of 9.6%, close to the currently recognized value of 9.5%. Similar preparations were obtained from nuclei of goose erythrocytes and other tissues. For some time, there was confusion between nucleic acid and chromatin, the nucleic acid-protein complex found in most cell nuclei, but by about 1890 the current distinction was clear.

THE CHEMICAL STRUCTURES OF NUCLEIC ACIDS

Although many were involved, the subsequent studies of the organic chemistry of nucleic acids were dominated by one figure: Phoebus Levene, of the Rockefeller Institute. Over the first three decades of the new century, Levene and his students largely established the fundamental structure of the units of

The Evolution of Molecular Biology. https://doi.org/10.1016/B978-0-12-812917-3.00006-1

deoxyribonucleic acid (DNA) as shown in Fig. 6.1. The repeating unit of the backbone of the DNA chain is a cyclic sugar, **deoxyribose**. A phosphate is attached to the 5′-hydroxyl of one unit to connect it to the 3′-hydroxyl of the adjacent unit. Repetition of this pattern forms the backbone of the DNA chain (Fig. 6.1A). One end of the chain, called the **5′-end**, will have an unreacted 5′-phosphate. The other end (**3′-end**) has unreacted 3′ hydroxyl group. Attached to each sugar is a **purine** or **pyrimidine base**, one of the four kinds designated A (adenine), T (thymine), G (guanine), and C (cytosine) (Fig. 6.1B). Thus, the whole molecule can be visualized as a necklace, with a repetitive chain decorated by regularly spaced charms. Levene also discovered the difference between two general classes of nucleic acids, previously known as yeast nucleic acid and thymus nucleic acid. The latter we now call deoxyribonucleic acid (DNA) because it lacks one oxygen on the ribose sugar of **ribonucleic acid (RNA)** (see the legend to Fig. 6.1). RNA also contains one base different from DNA—uracil (U) rather than thymine (T). These differences may seem small, but we shall see that they are of major importance in determining the different functions of these two substances.

Although Levene had almost single-handedly elucidated the chemical structure of the nucleic acids, he was also largely responsible for one of those sidetracks that are so common in science—the **tetranucleotide hypothesis**.

It must be noted that as early as the last decade of the 19th century, the work of Miescher and his school had suggested to many scientists that "Nucleïn" was the material that transmitted genetic information. But this came in doubt as the basic chemistry of DNA was unraveled. The fact that there were four bases in approximately equal quantities led Levene and other to espouse the idea that DNA was composed of aggregates of "tetranucleotides" each containing one each of the four bases. It must be remembered that the notion of long-chain polymers was unacceptable to most organic chemists of that period (see Chapter 3 for comparable attitudes on protein chains). The tetranucleotide hypothesis carried the corollary that such a simple structure could not carry genetic information—that must be left to the more complex proteins. Nucleic acids, in this view, were some kind of structural material in the nucleus—perhaps binding the hypothetical protein genes into chromosomes. This wholly incorrect view persisted until nearly the mid-20th century, reinforced by the enormous prestige of Phoebus Levene. Even after the macromolecular nature of nucleic acids became recognized, the concept of their function as structural materials persisted.

"WHAT IS LIFE?"

A remarkable clarification of the necessary features of the hereditary substance came from an unexpected source. In 1943, Erwin Schrödinger, one of the legendary founders of quantum mechanics, was asked to give a series of lectures, subject of his choosing, at the University of Dublin, Ireland. Why

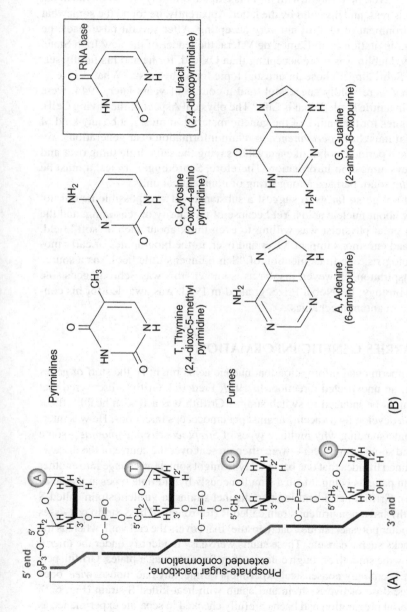

FIG. 6.1 Chemistry of nucleic acids. (A) A tetranucleotide. The backbone of the chain is formed by the phosphates and sugars; the bases are indicated by circled letters. Note that the chain possesses polarity: the 5′ end and the 3′ end are chemically different. Reading the sequence from the top, in 5′ to 3′ direction, will give ATCG; reading from bottom, in 3′ to 5′ direction, will give GCTA. By convention, nucleotide sequences in nucleic acids are written and read from 5′ to 3′. Since nucleotide units all share phosphates and sugars and differ in only in the purine or pyrimidine base attached, the nucleotides carry the name of the attached base. (B) Chemical formulas of pyrimidines and purines. Uracil (boxed) is only present in RNA.

Schrödinger was in Dublin is a tale in itself. He had fled Germany to escape Fascism, as had so many European scientists, and initially accepted a post at Oxford. However, he brought with him his extended family which included his wife, his mistress, and his child by the latter. Apparently, he found the somewhat stuffy environment of Oxford not very accepting. After several false leads, he accepted an invitation from Éamon de Valera, the leader of the new Irish State. Apparently, Dublin was more accepting than Oxford, for he and his family settled there. Schrödinger chose an unusual topic for a physicist: "What is Life?". The lectures were wildly successful, and a couple of years later, 1944, were published in a little book (What is Life? The physical Aspect of the Living Cell). In it, he argues for the nature of the genetic material: it must not be any kind of liquid, for it must have *permanence* to retain information over generations. Yet it cannot be a periodic solid, like a crystal; saying the same little thing over and over conveys almost no information. Therefore, Schrödinger argues, it must be an *aperiodic* solid, perhaps a long string of nonidentical units.

He did not go so far as to suggest a substance—it was possible he did not even know about nucleic acids. Yet because of the clarity of reasoning, and the fact that a great physicist was willing to even think about such a "soft" field, the book had enormous impact. Over and over, in the biographies of early molecular biologists, one finds mention of "Schrödinger's little book" as a source of early inspiration. However, insofar as is known, this was Schrödinger's sole foray into biology: his Nobel Prize, granted in 1933, was awarded for his contribution to quantum mechanics.

DNA CARRIES GENETIC INFORMATION

The first experimental information that nucleic acids might be the stuff of genes came from an unexpected direction. In 1927, Frederick Griffith discovered that bacteria might be induced to switch strains. Griffith was a British health officer working on developing a vaccine against pneumococcus infections. He was interested in understanding why multiple types of *Streptococcus pneumoniae*—some virulent and some nonvirulent—were often present over the course of the disease. He entertained the idea that one bacterial type might somehow change into another (rather than patients being infected simultaneously by multiple types at the onset of disease). Griffith had identified two distinct strains: a virulent strain called S (for smooth) and a nonvirulent strain called R (for rough). The S-strain possesses a smooth outer polysaccharide coat (capsule) that covers the cell wall, whereas the R-strain lacks such a capsule. These stains were easy to identify under the microscope, or by the smooth or rough colonies they formed on agar plates. Griffith performed the seemingly nonsensical experiment of injecting one mouse *twice*; once with a large dose of live R-strain and again with heat-killed S-strain (Fig. 6.2). Each bacterial preparation had been carefully checked in separate experiments; as expected, neither was lethal. Yet in many cases, the mouse died! And it contained living type S bacteria! Somehow, the live R has transformed into S!

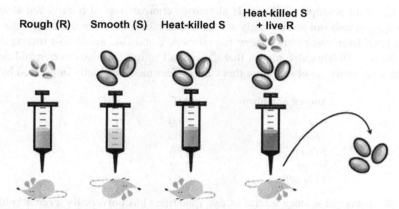

FIG. 6.2   The Griffith experiment. After the mouse is injected with heat-killed S and live R, live S bacteria are isolated from the tissues of the dead mouse. Something must have "transformed" the live R into live S.

Griffith's experiment intrigued Oswald Avery, a researcher at the Rockefeller Institute in New York, and he and his students began to search for the substance responsible for this "transformation" of bacteria. The search was long, well over a decade. They first found that no mouse was needed; transformation would happen even if living R-strain was mixed in the test tube with heat-killed S. The S could be fragmented as well, but the mix of its fragments would still induce the change. As more and more substances were eliminated from the list of cell constituents that could be the transforming "principle," one persisted—DNA. Finally, highly purified DNA from S-strain would suffice. In 1944, Avery and co-workers announced their results in a modest paper in the Journal of Experimental Medicine. There was no press conference, no claim that DNA was the stuff of genes. The paper clearly showed that DNA, by itself, can transmit information to a cell and change its phenotype. The paper, like Griffith's, was largely ignored at the time. Yet, the course of biology had been nudged in a new direction.

MYSTERIOUS NUMBERS

By 1950, the chemical structure of DNA was well understood, and its possible biological significance appreciated by many. About this time, a very experienced organic chemist at Columbia University began examining the comparative base compositions of a variety of DNA from different sources. The chemist was Erwin Chargaff, an expatriate from Austria, and a very learned man, in the classics as well as organic chemistry—he liked to sprinkle his writing with obscure Latin phrases and references to Greek mythology. He was also the master of truly withering sarcasm, as we shall see in the next chapter.

Chargaff cleaved the polynucleotides with alkali, separated the products by the newly developed chromatographic methods, and measured their amounts by UV light absorption. He made numerous comparisons of relative amounts, but what stands out so strikingly are the molar ratios shown in the table below, extracted from one of his papers: the ratios A/T and G/C are almost invariably close to 1.00! This also means that the overall ratio of purines to pyrimidines is close to unity, an observation that Chargaff seemed especially impressed by.

Source organism	A/T	G/C
Human	1.00	1.00
Chicken	1.00	1.00
Salmon	1.06	0.91
Wheat	1.02	1.02
Escherichia coli, K12	1.09	0.99

Whenever in science a ratio of two quantities has universally a remarkable ratio like 1.00, it should be cause for the most avid attention. This *says* something. Nowadays the significance is blazingly clear: DNA is made of two complementary strands, with A pairing with T and G pairing with C. Yet somehow, neither Chargaff nor his contemporaries tumbled to the fact. For several years, Chargaff's ratios were just mysterious numbers.

EPILOGUE

In the first half of the 20th century, nucleic acids passed from obscurity in chemical structure and biological function to being recognized as well-defined and possibly vital molecules in genetics. All of the momentum from previously diverse and unconnected channels of research now focused on the questions: what is the structure of DNA and what will it tell us? Yet, as late as 1950 neither seemed close to an answer. That had to come from a whole new generation of scientists with remarkably diverse backgrounds.

FURTHER READING

Books and Reviews

Hausman, R.E., 2002. To Grasp the Essence of Life. Kluwer Academic Publishers, Dordrecht, The Netherlands. Excellent discussion of Griffith, Avery, etc.

Judson, H.F., 1979. The Eighth Day of Creation. Makers of the Revolution of Biology. Simon & Schuster, New York. A commemorative edition by Cold Spring Harbor Laboratory Press, Cold Spring Harbor, NY, 2013. Enormous detail on matters in this chapter and the several to follow. A Source!

Moore, W.J., 1994. A Life of Erwin Schrödinger. Cambridge University Press, Cambridge, UK. A vividly written account of Schrödinger's life.

Schrödinger, E., 1944. What is Life? The Physical Aspect of the Living Cell. Cambridge University Press, New York, NY. The classic "little book" that influenced so many. Much good here even today!

Zlatanova, J., van Holde, K.E., 2016. Molecular Biology. Structure and Dynamics of Genomes and Proteomes. Garland Science. Taylor & Francis, New York, NY. Much more detail on proteins/nucleic acids and methods used for their study will be found in this comprehensive textbook.

Classic Research Papers

Avery, O.T., MacLeod, C.M., McCarthy, M., 1944. Studies on the chemical nature of the substance inducing transformation of pneumococcal types: Induction of transformation by a desoxyribonucleic acid fraction isolated from Pneumococcus type III. J. Exp. Med. 79, 137–158. Description of the major contribution of Avery and his group.

Chargaff, E., 1951. Structure and function of nucleic acids as cell constituents. Fed. Proc. 10, 654–659. This paper contains Chargaff's important results on ratios in DNA.

Griffith, F., 1928. The Significance of Pneumococcal Types. J. Hyg. (London) 27, 113–159. The paper that stimulated Avery's research.

Chapter 7

The Great Synthesis

PROLOGUE

By the late 1940s, it had become clear that DNA was the material from which genes were made, and that it somehow coded the information for the construction of proteins. Yet, the synthesis between genetics and biochemistry was still not complete. Both sciences had followed their individual paths to ever greater complexity, but this seemed to hinder, rather than aid in making connections. Much of genetics had concentrated on complex animals like fruit flies. On the other hand, the structural details of nucleic acids and proteins seemed inaccessible. Simpler systems and new methods were needed. As is often the case, these breakthroughs were found by a cadre of young scientists, many from fields diverse from genetics or chemistry.

DO BACTERIA AND BACTERIOPHAGE HAVE GENETICS?

In 1937, a young physicist named Max Delbrück came to The California Institute of Technology on a Rockefeller Institute Fellowship. Surprisingly, his intent was to work with Thomas Hunt Morgan on genetics. Delbrück had an interesting background: descended from an academically distinguished German family, he had shown great promise in physics, and had attracted the attention of the great Danish physicist Niels Bohr, with whom he had studied. Although his actual accomplishments to this point were few, Delbrück was known in the European physics community for his sharp mind and acute criticisms of sloppy work. It was under Bohr's tutelage that Delbrück had gained an interest in biology, for Bohr felt that biological processes could be explained in physical terms. Delbrück had coauthored an important paper on the production of mutants by high-energy radiation, following on the work of Muller (see Chapter 2). It was this paper that inspired Schrödinger to write "What is Life."

Delbrück soon realized that the genetics of fruit flies was not the entree for a physicist into genetics. It was just far too complicated, and unsuitable for simple, quantitative experiments. Fortunately, there was at Caltech another young scientist, Emery Ellis, who was doing pioneering studies on bacteriophage, or "phage," as they came to be called. Bacteriophage are viruses that attack bacteria. Ellis and Delbrück soon devised a very simple experimental protocol that revealed the three stages in phage infection: (1) attachment to and entry into the bacterium, (2) replication of the phage therein, and (3) lysis of the bacterium and release of new phage (Fig. 7.1). The genetics of phage and bacteria had long

The Evolution of Molecular Biology. https://doi.org/10.1016/B978-0-12-812917-3.00007-3

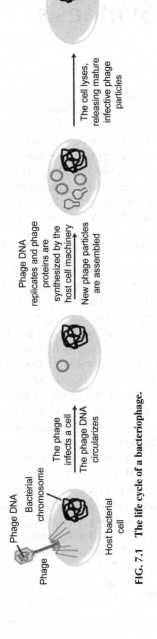

FIG. 7.1 The life cycle of a bacteriophage.

been neglected, for neither follows the laws of classical Mendelian genetics. Both are haploid, in other words, the genetic material is present in one copy, under most circumstances. Nevertheless, the simplicity of the system allowed the kinds of experiments physicists like to do. At that point in time, bacteriophage were the ideal model system. Many advances in biology have depended upon finding just the appropriate creature as a model (see Box 7.1).

BOX 7.1 Model Organisms

Throughout this book, we have several times noted that the use of a particular organism was especially appropriate for a given study. In addition, certain organisms have been used again and again as convenient models for whole categories of organisms (Figure 1). A brief description of some of these model organisms follows below, with notes as to why they have been so often chosen. The genomic sequences of all of these organisms are now available.

Today, bacteriophage *lambda* is employed largely as a cloning vector but it played an important part in the early development of genetics, especially because it has two alternative life cycles: lytic and lysogenic (see above).

The bacterium *Escherichia coli* has been labeled the "workhorse of molecular biology." There is practically no fundamental biochemical process, from DNA replication to protein synthesis, that was not first elucidated in *E. coli*. It is extremely easy to grow, in liquid culture or on solid agar plates, and metabolically very versatile, which has made it useful for studies of metabolic regulation.

The common budding (baker's) yeast *Saccharomyces cerevisiae* is among the simplest eukaryotes. It is unicellular, thus providing a bridge between the bacteria and the more complex eukaryotes. Its genetics has been very thoroughly studied, with many knockout strains available. In a knockout strain, a particular gene is inactivated by recombinant DNA techniques (Chapter 14). Studying such knockouts helps in elucidating the biological functions of genes. One difficulty in working with *S. cerevisiae* is the tough outer cell wall, which makes it difficult to insert substances. The "fission yeast" *Schizosaccharomyces pombe* is genetically similar to *S. cerevisiae* but lacks the tough integument. It divides, rather than buds, which is an advantage for some studies.

The free-living, primitive, unsegmented, and bilaterally symmetrical worm *C. elegans*, which was introduced to the field by Sidney Brenner, is a remarkably simple creature. The adult worm has only 1,090 cells, the lineage of each of which is precisely known (Figure 2). This makes it an outstanding candidate for developmental studies, as recognized by the 2002 Nobel Prize in Physiology or Medicine awarded to Sidney Brenner, Robert Horvitz, and John Sulston.

Drosophila melanogaster was the organism which provided the seminal studies in modern genetics. Morgan's fruit fly is easy to grow, in enormous numbers, in a very short time. This, plus the availability of a great many mutant strains, including many with mutations that affect general developmental patterns, make it still a useful model. The embryos are also used, especially for biochemical studies.

Continued

BOX 7.1 Model Organisms—Cont'd

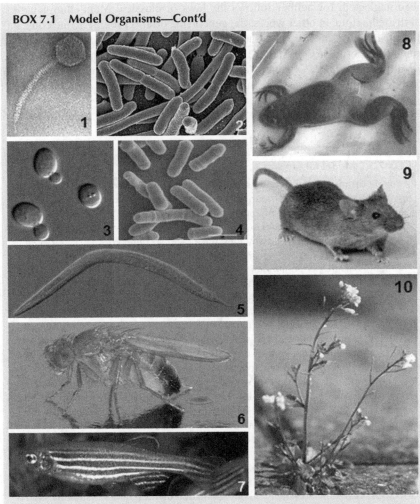

FIGURE 1. A picture gallery of some of the model organisms used most frequently in genetics research. *(Images from Wikimedia and Wikipedia.)*

1. λ phage of *Escherichia coli*
2. *Escherichia coli*
3. *Saccharomyces cerevisiae*
4. *Schizosaccharomyces pombe*
5. *Caenorhabditis elegans*
6. *Drosophila melanogaster*
7. *Danio rerio*
8. *Xenopus laevis*
9. *Mus musculus*
10. *Arabidopsis thaliana*

BOX 7.1   Model Organisms—Cont'd

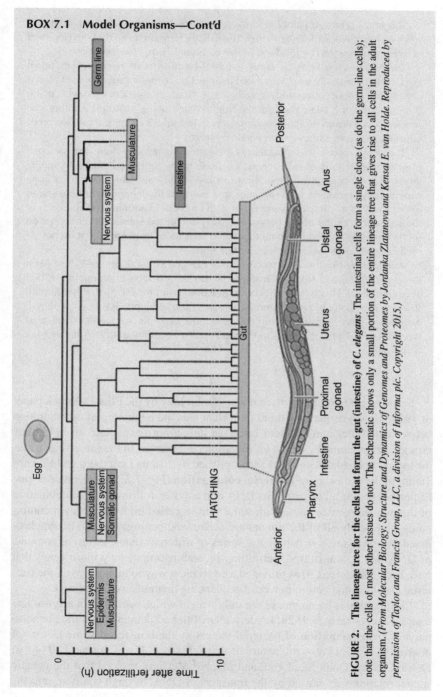

FIGURE 2. The lineage tree for the cells that form the gut (intestine) of *C. elegans*. The intestinal cells form a single clone (as do the germ-line cells); note that the cells of most other tissues do not. The schematic shows only a small portion of the entire lineage tree that gives rise to all cells in the adult organism. (*From Molecular Biology: Structure and Dynamics of Genomes and Proteomes by Jordanka Zlatanova and Kensal E. van Holde. Reproduced by permission of Taylor and Francis Group, LLC, a division of Informa plc. Copyright 2015.*)

Continued

The small zebrafish *Danio rerio* is easy to grow and is very fecund, so it provides a convenient vertebrate model. The fish are transparent so that development of internal organs can be followed in live embryonic fish. How convenient!

The African frog *Xenopus laevis* is useful because of its large and abundant eggs, which can easily be manipulated by injection of foreign substances. *X. laevis* can rapidly produce thousands of embryos. A disadvantage of this model is that it is tetraploid and it takes years for the frog to reach sexual maturity. Another frog species, *X. tropicalis*, is diploid and matures in less than 3 months; these two properties make it very attractive for genetic research.

The house mouse *Mus musculus* is the easiest mammal to study and has been used by generations of researchers. By now, many pure-bred strains including those with specific genetic modifications are readily available. Despite a large evolutionary separation between mice and humans, the laboratory mouse shares the majority of its protein-coding genes with humans. In addition, comparisons of the sequences of the entire genomes (2014) confirmed substantial conservation in the newly annotated potential functional sequences (sequences that do not directly encode proteins but have other functions).

Arabidopsis thaliana is a weed commonly known as thale cress. The plant is easy to grow and sexually matures in less than 6 weeks, producing ~5000 seeds per plant! This is the most commonly used plant model. First, it has a small genome (five pairs of chromosomes) whose entire sequence has been reported. Second, a large number of mutant lines and genomic resources (databases) are available. Third, it is easy to transform using recombinant DNA technology techniques.

Some might mark the birth of molecular biology by the Ellis-Delbruck paper of 1939. However, perhaps more important was the beginning of collaboration between Delbrück and Salvador Luria, an Italian-born expatriate. In 1943, they provided convincing evidence for mutations in bacteria. In the same year, a major technical breakthrough was accomplished by Joshua Lederberg and Edward Tatum in the discovery of **bacterial conjugation** (Fig. 7.2). In this process, one bacterium inserts all or part of its DNA into another, followed by recombination of the two DNA molecules. With some strains, called high-frequency recombinants, practically all of the donors are active, and conjugation can be synchronized. If conjugation is halted at a series of different times, different amounts of DNA will be transferred, permitting recombination of only those genes that have been transferred. This provided a convenient way to map genes on the bacterial chromosome before powerful sequencing methods were available.

Conjugation is by no means the only way in which bacteria can acquire foreign DNA. As early as 1928, Frederick Griffith had demonstrated the phenomenon of **transformation** of bacterial strains by the transfer of some factor. Of course, he did not know the nature of that substance. It was not until 1944 that Oswald Avery, Colin MacLeod, and Maclyn McCarty showed that the genetic transformation was caused by the transport of DNA, through solution, into the

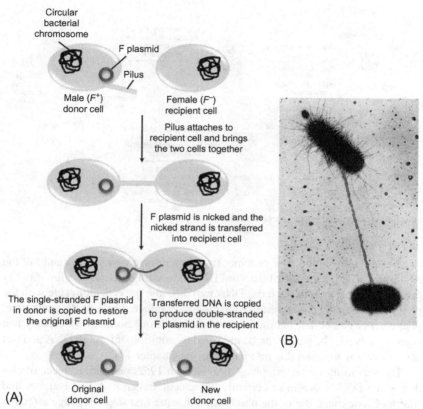

FIG. 7.2   Bacterial conjugation. Conjugation between two bacterial cells can occur only when one of the partners carries the so-called F (fertility) plasmid; these cells are known as F-positive or F-plus cells. The F plasmid exists as an episome, independently of the main bacterial chromosome. It carries its own origin of replication, an origin of transfer (where nicking occurs to initiate transfer to a recipient F- cell), and a whole battery of genes responsible for the formation of the pilus and attachment to the recipient cell. (A) The steps in the process. (B) Electron microscopic view of two bacterial cells in the process of conjugation. *(From http://www.flickr.com/photos/ajc1/1103490291; credit: Alan Cann, alan.cann@leicester.ac.uk.)*

transformed bacterium. These experiments are at the very basis of molecular biology, as described in detail in Chapter 6.

Finally, DNA can be transferred between bacteria via bacteriophage, by a process called **transduction** (Fig. 7.3). This usually involves what are known as **temperate phage**, which can exhibit two alternative life cycles: lytic and lysogenic. In the lytic cycle the virus can enter the bacterial cell, replicate, lyse the host bacterium, and go on to infect other bacteria. In the alternative (lysogenic) cycle, it can integrate its DNA into the bacterial chromosome. The viral DNA may remain in the bacterial chromosome in a dormant state for many bacterial generations, until some stimulus (such as radiation or chemical insult) causes it

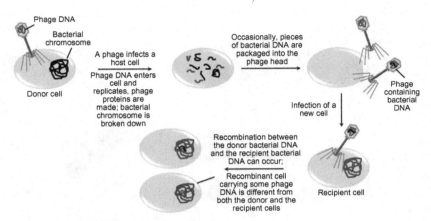

FIG. 7.3   Conceptual schematic of transduction.

to be released from the host genome. It will then form new viruses and kill the bacterium. The packaging of the viral DNA into viral particles is a low-fidelity process: small pieces of bacterial DNA may become packed, alongside with the phage genome, and thus transferred to the newly infected cell. At the same time, phage genes can be left behind in the bacterial chromosome in which they had been integrated. The phage can be modified by addition of foreign DNA and can act as a **vector** to insert this into bacteria (see Chapter 14).

The versatility of bacteriophage provided, in 1952, one further piece of evidence that DNA was almost certainly the genetic material. Alfred Hershey and Martha Chase made use of the observation that the first stage in phage infection of a bacterium involved attachment of a fragile tail to the bacterium, followed by transfer of *something* into the microbe that allowed generation of new phage. But was it protein or DNA? They generated one group of phage with radioactive phosphorus, which would label almost exclusively the DNA. Another batch of phage had radioactive sulfur, to label the phage protein. Each phage sample was then mixed with bacteria and given time to carry out the insertion of the critical material. Then a simple kitchen blender was used to shear off the attached phage, and the product centrifuged. The bacteria at the bottom of the tube were separated from remaining phage fragments in the supernatant. Most of the radio phosphorus (and hence DNA) was with the bacteria; most of the labeled protein stayed outside (Fig. 7.4). Thus, the material that phage inserted into bacteria for replication and generation of new phage was DNA. In a sense, the "blender experiment" confirmed Avery's work of a decade earlier (Chapter 6), but it was much more direct and specific. By 1952, only a few would doubt that DNA was the genetic material.

Luria, Delbrück, and the group of young scientists who began to work with bacteria and phage have become known as the "phage group." Some of the

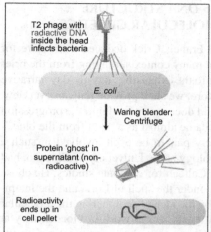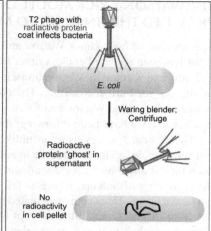

FIG. 7.4   The Hershey-Chase experiment.

TABLE 7.1   Early members of the "phage group"

Name	Year of birth	Country of birth	Field of training
Crick, Francis*	1916	United Kingdom	Physics
Delbrück, Max*	1906	Germany	Physics
Gamow, George*	1904	Russia	Physics
Hershey, Alfred*	1906	United States	Microbiology
Lederberg, Joshua*	1925	United States	Medicine
Luria, Salvador*	1912	Italy	Medicine
Szilard, Leo	1898	Hungary	Physics
Watson, James*	1928	United States	Biology

Asterisk denotes scientist who won the Nobel Prize for their work (see Appendix).

principals are listed in Table 7.1. The group was truly international in origin, even if most worked in the United States. What is remarkable is the diversity of backgrounds, with little representation of biochemistry or genetics. The multiple ways in which the DNA of bacteria and bacteriophage can be readily manipulated explains why the phage group played such an important role in the early development of molecular biology. But even as late as 1952, the cornerstone of the whole field was still missing: the structure of DNA.

THE WATSON-CRICK MODEL OF DNA STRUCTURE PROVIDED THE FINAL KEY TO MOLECULAR GENETICS

The account of how James Watson and Francis Crick deduced a structure for DNA has been repeated endless times in many contexts, ranging from the brief and informal account by Watson himself to the exhaustive, day-by-day narrative by Judson (see Further Reading). Therefore, we shall provide only an overview. In many ways, what Watson and Crick did does not fit well into the progression of ideas that we have been following; it came almost as a "bolt from the blue."

The principal actors were an unlikely pair to be even involved in such a project. Watson, an undergraduate in zoology at the University of Chicago, had been turned down by both Harvard and Caltech for graduate studies. He chose the University of Indiana, where he fell under the spell of Luria and the incipient phage group. He became fascinated by the idea of deciphering the molecular structure of genes. An initial attempt at postdoctoral work in biochemistry at Copenhagen had proved uninteresting. However, a chance encounter with the X-ray diffraction studies being carried out in Cambridge University impelled him to transfer his fellowship there. He was in his early twenties, and at least in a formal sense, wholly unequipped for what he planned to do.

Watson soon made the acquaintance of Francis Crick, a physicist by training, who was returning to Cambridge to finish his PhD, after serving several years in military research during the war. Crick had the same goal, but no background in biology. Interestingly, both Watson and Crick claim that their interest was inspired by Schrödinger's little book.

The prospects did not seem bright. Whereas the early work of Astbury (Chapter 4) had demonstrated that X-ray diffraction from fibers formed by macromolecules had shown the feasibility of obtaining structural information, Watson and Crick initially had no such data. Neither were they likely to obtain data themselves, for they had no experience in the experimental technique, nor access to the apparatus with which to obtain it. Their inexperience horrified Chargaff when he visited them in 1952—for example, they were uncertain of some of the purine and pyrimidine structures. Chargaff says "I never met two men who knew so little, and aspired to so much." He says that he described his DNA composition studies to them, of which they seemed ignorant.

It should be pointed out that the inexperience of Watson and Crick was not necessarily a handicap; it often happens in science that the investigator naive to a field can have insights inaccessible to the experts. His or her view is not cluttered by all the details from previous work. Note the backgrounds and accomplishments of the phage group (Table 7.1).

In essence, Watson and Crick must have decided to mimic Pauling's remarkable deduction of the protein α-helix, doing modeling constrained by known parameters, gleaned from a variety of sources. They presumed that the deoxyribose rings would be rigid, with chain flexibility only in the phosphodiester links between them. This backbone structure had been recently elucidated by a renowned organic chemist, Lord Alexander Todd. They knew from even the earliest diffraction patterns available to them that the structure must be helical, with a regular spacing of about 0.34 nm along the axis. The known density of DNA was too great to allow a single-strand helix; it must involve two or more strands. That was all they knew. And they were not alone in their quest for the structure. At Caltech, the formidable Linus Pauling was hard at work on model structures; this was fearful competition. At King's College, in London, in the laboratory of Maurice Wilkins, a young woman named Rosalind Franklin was obtaining the best X-ray diffraction patterns from DNA that had been produced to date. No collaboration existed between the laboratories, and Franklin was apparently not overly interested in the enormous biological import of DNA, she regarded it as a challenging structural problem.

Competition from Pauling turned out to be of no significance. The great man proposed a three-strand helix model, with the sugar-phosphate backbones crammed into the center of the helix and the bases projecting outward into the solution. The latter seemed eminently reasonable—if DNA is a "tape" to be read in terms of base sequence, those bases should be outside where they are accessible to reading. But the Pauling model required suspiciously close packing of charged phosphates, did not account for Chargaff's ratios, and did not suggest a mechanism for replication. Even before its publication, Watson and Crick had dismissed it, for they had obtained a copy of the manuscript from Linus Pauling's son Peter, who had just arrived to study at Cambridge.

Critical to Watson and Crick's further progress was the high-quality X-ray diffraction data obtained by Franklin. They obtained this information as part of a report by Franklin to the Medical Research Council. The report was by no means secret, but it put the critical data on the parameters of the helix (base spacing, helical repeat, number of units per turn of the helix, and diameter of the helix) in the hands of two who had contributed none of those data. With this information, they could begin to build realistic models. The big problem was where to put the purine and pyrimidine bases. Details of the diffraction pattern indicated two strands, and indicated that the relatively massive phosphate-ribose backbones must be on the outside, leaving the bases in the center of the double helix. However, if simply arranged in random sequences on the two strands, the bases did not fit well, and gave a lumpy helix. While waiting for wire models to be constructed, Watson began playing with cardboard cut-outs. Suddenly, on a late February day in 1953 he noticed that A and T, as well as G and C could fit together, connected by hydrogen bonds! The structure, as schematized in Fig. 7.5A, came almost automatically. Furthermore, the base pairing explained Chargaff's rules, and the distance between the two backbone chains was identical for the two allowed pairs (Fig. 7.5B). Thus, a smooth double helix of 2.0 nm diameter would result.

Announcement of the proposed structure came in a very short note published in the journal Nature in April 1953. It is only a little over a page in length and provides no fine details of the structure. It says nothing about the function

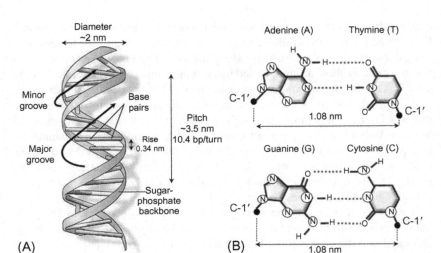

FIG. 7.5   Schematic of the double-helix structure of DNA in the B-form. (A) Base pairing occurs between the purines on one strand and the pyrimidines on the other. (B) The pairing between these "complementary" bases allows the distances between the C-1′ on the sugar moieties to be exactly the same for both adenine–thymine and guanine–cytosine base pairings. The H-bonding between the complementary bases and the "stacking" interactions between successive pairs are both important in stabilizing the helix. The original Watson–Crick model closely resembles this, but with 10.0 base pairs per turn. *(Adapted from the U.S. National Library of Medicine).*

of DNA, although it does coyly note that the complementary chain structure suggests a possible self-copying mechanism. Yet, scant as it is, it marks perhaps the greatest single step in the history of biology. Surprisingly, the paper did not receive great attention at first. Perhaps the clearest recognition came from Delbrück, the first person Watson told of the model. Delbrück is said to have remarked: "One cannot imagine that nature would not have made use of such a phenomenal discovery."

The discovery earned the Nobel Prize for Watson, Crick, and Wilkins in 1962. We can never know whether or not Franklin would have shared, had she not died of cancer in 1958. The Nobel prizes are awarded only to living scientists.

EPILOGUE

By 1955, two years after Watson and Crick's publication of the DNA double-helix model, the molecular basis of genetics was becoming clear. Genes are made of DNA, which is carried in chromosomes. Bacteria and phage have haploid chromosomes, constituted of one or a few double-helical DNA molecules. These simple systems provided the entree into molecular genetics, and key ideas for a molecular biology. Mutations occur by modification of DNA sequences, and exchange of alleles occurs by recombination. However, almost none of the details or control of these processes was understood in 1955. Much has been learned since then, as we shall show in following chapters.

FURTHER READING

Books and Reviews

Delbrück, M., 1949. A physicist looks at biology. Transactions of the Connecticut Academy of Arts and Sciences 38, 173–190. A view of molecular biology from one of the pioneers.

Judson, H.F., 1979. The eighth day of creation. Makers of the Revolution of Biology. Simon & Schuster, New York. A commemorative edition by Cold Spring Harbor Laboratory Press, Cold Spring Harbor, NY, 2013. Exhaustive detail on the early days of molecular biology, almost a day-by-day account!

Lederberg, J., 1948. Problems in microbial genetics. Heredity (Edinburgh) 2, 145–198. From one of the pioneers in the field.

Brenner, S., 1974. The genetics of *Caenorhabditis elegans*. Genetics 77, 71–94. Covers the use of mutations in the worm to answer questions about development and differentiation, focusing on the nervous system.

Luria, S.E., 1947. Recent advances in bacterial genetics. Bacteriological Reviews 11, 1–40. A summary of results of bacteriological experimentation which satisfy the quantitative requirements of modem geneticists.

Watson, J., 1968. The Double Helix. A Personal Account of the Discovery of the Structure of DNA. Atheneum, New York, NY. A personal and somewhat idiosyncratic view of the events.

Classic Research Papers

Ellis, E., Delbrück, M., 1939. The growth of bacteriophage. Journal of General Physiology 22, 365–384. Delbrück's first adventures with phage.

Lederberg, J., Tatum, E.L., 1946. Gene recombination in *Escherichia coli*. Nature 158, 558. Describes the use of mutant bacterial strains lacking the ability to synthesize specific growth factors; these mutants will only grow in media supplemented with nutrients, and thus can be selected for.

Tatum, E.L., Lederberg, J., 1947. Gene recombination in the bacterium *Escherichia coli*. Journal of Bacteriology 53, 673–684. Presentation of evidence for the occurrence in a bacterium of a process of gene recombination, akin to a sexual stage in the life of bacteria.

Chapter 8

How DNA is Replicated

PROLOGUE

"It has not escaped our notice that the specific pairing we have postulated immediately suggests a possible copying mechanism for the genetic material." This
cryptic sentence, inserted near the end of Watson and Crick's note, has often
been criticized as overly coy. However, one cannot fault them for inserting it,
for its statement is certainly true, and if it had not been there, some fame-seeker
would have jumped on the opportunity to claim it. However, things did not turn
out to be nearly as simple as this statement implied, and it took decades to fully
unravel the story. There were a number of difficult questions to be asked.

WHAT IS THE MODE OF REPLICATION?

As Fig. 8.1 shows, there are several ways in which one could imagine going
from one double helix (duplex) to two. In **conservative replication** (Fig. 8.1A),
some mechanism copies both strands to make an all-new double-strand molecule. **Semiconservative replication** (Fig. 8.1B) requires splitting the original
duplex and forming a complimentary copy of each strand. Here, each product duplex would be half-old, half-new DNA. Finally, **dispersive replication**
(Fig. 8.1C) could result if there were mixing of portions of chains. Which was
the case *in vivo*?

  The question was answered by a simple, elegant experiment carried out by
Matthew Meselson and Frank Stahl in 1957. They used a new technique called
density gradient centrifugation, in which a solution of a dense salt like cesium
chloride (CsCl) is centrifuged long enough to set up a stable density gradient
in the solution. A macromolecule dissolved in that solution will migrate to the
point corresponding to its own density and form a band there. What Meselson
and Stahl realized is that the method is exquisitely sensitive to differences in
density. It could easily separate DNA molecules labeled with nitrogen isotope
14 (^{14}N) from those with the slightly heavier isotope ^{15}N, even though these
differ in density by only 0.014 g/mL. A schematic of the results is shown in
Fig. 8.2. Bacteria (*Escherichia coli*) were first grown in a medium where the
only nitrogen source contained the heavy isotope ^{15}N. DNA from these bacteria banded at the density marked ^{15}N. The bacteria were then transferred to a
medium containing only ^{14}N and allowed to reproduce. After one generation,

The Evolution of Molecular Biology. https://doi.org/10.1016/B978-0-12-812917-3.00008-5

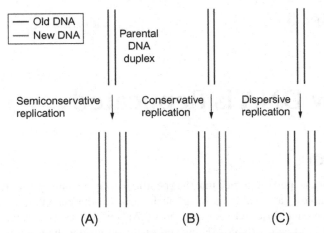

FIG. 8.1 Three modes of DNA replication that were considered as possibilities.

the DNA band had moved to a position halfway between ^{15}N and ^{14}N. This immediately argued for the semiconservative mode: all duplexes were now half-light and half-heavy DNA! Behavior after further generations confirmed this; for example, after two replications in the ^{14}N medium, Meselson and Stahl observed one light-light band and one light-heavy hybrid. This also rules out dispersive replication. What Meselson and Stahl did has been called by some "the most beautiful experiment." Beautiful it was, but not entirely original. As early as 1941, the eminent biologist John B.S. Haldane had suggested the use of nitrogen isotopes to distinguish new genes from old. The technique that made that possible, density gradient sedimentation equilibrium, was conceived and developed by an American physical chemist, Jerome Vinograd. He has rarely received the credit deserved.

HOW DOES REPLICATION PROCEED?

Clear as it was, the Meselson-Stahl experiment told nothing about the actual mechanism of replication. The first evidence for this came from a young bio-chemist named Arthur Kornberg at Washington University, Saint Louis. Using radiolabeled nucleotides, he demonstrated that these would add to a polynucleo-tide chain in the 5′ to 3′ direction, in the presence of template DNA (Fig. 8.3). Kornberg also identified an enzyme, which he termed **polymerase**, that cata-lyzed the reaction. Kornberg was awarded the 1959 Nobel Prize in Physiology or Medicine (shared with Severo Ochoa) for "discovery of the mechanisms in the biological synthesis of ribonucleic acid and deoxyribonucleic acid."

The fact that the polymerization reaction proceeded most efficiently with the nucleoside triphosphates (instead of di- or monophosphates) resolved another question—where did the energy to drive the process come from? It was already

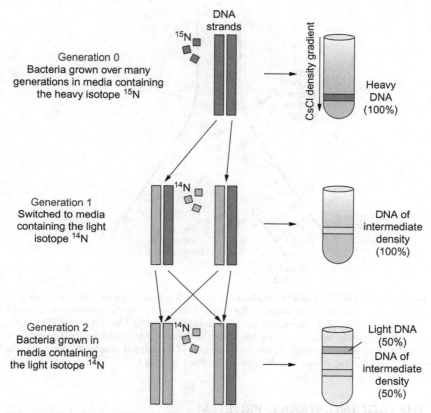

FIG. 8.2 The Meselson Stahl experiment, establishing the semiconservative mode of DNA replication.

known that hydrolysis of phosphate groups off adenosine triphosphate was widely used as a cellular energy source, driving processes as diverse as chemical reactions, muscle contraction, and transport through membranes. It is in the enzymatic synthesis of DNA that genetics first encountered biochemistry. The fundamental process in reproducing organisms depends on the kind of catalytic processes and energy sources that had intrigued biochemists for over a century.

The picture seemed complete until a disturbing experiment was carried out by John Cairns and Hector DeLuca in 1969. They found that an *E. coli* mutant, devoid of the Kornberg enzyme, lived happily! The mutant was a little more sensitive to radiation but replicated its DNA normally. This could only mean that some other enzyme catalyzed the replication of the *E. coli* genome. A search was undertaken. In fact, *two* new enzymes were found. Of the three, now named **Pol I**, **Pol II**, and **Pol III**, only pol III was necessary for cell division. The Kornberg enzyme (now Pol I) and Pol II played subsidiary roles in the cell (see later).

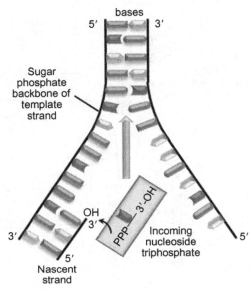

FIG. 8.3 The chemistry of DNA replication. The DNA polymerase at the replication fork uses nucleoside triphosphates as substrates for the polymerization reaction; the nucleotides to be added to the nascent DNA chain are selected according to the base-pairing rules that govern DNA structure. The antiparallel directions of the two strands in the double helix and the requirement for a free OH group allow continuous synthesis on only one of the parental strands; the other strand is synthesized by a discontinuous mechanism, as illustrated in Fig. 8.4.

THE LAGGING-STRAND PROBLEM

There was still one more puzzle to face. All nucleotide polymerases that have ever been studied elongate the chain in the 5′ to 3′ direction. That is, they track along the complementary template chain going from its 3′ end toward its 5′ end. But as Fig. 8.4 shows, this creates an awkward situation. One strand (called the **leading strand**) can be continuously generated. But what about the other, the so-called **lagging strand**? One might have thought nature would have evolved a polymerase working in the opposite direction, but that is not so. Rather, there evolved a rather byzantine mechanism, in which short segments of complementary DNA are generated on the lagging strand. Polymerization is in the 5′ to 3′ direction, just as in the leading strand, and Pol III and Pol I are both involved. This direction means that the polymerization on the lagging strand must be started numerous times during the replication process, and *that* uses RNA for short **primers**, which must later be removed. Then the gaps are filled and the pieces ligated together. This requires additional enzymes: a **primase**, a **DNA ligase**, which ligates the Okazaki fragments (see below) together in an uninterrupted chain, and a **DNA gyrase**, which facilitates the unwinding of the DNA to separate the leading and lagging strands (Fig. 8.5). It all seems unbelievably

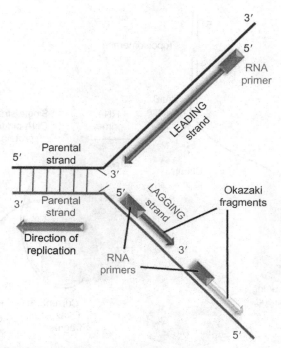

FIG. 8.4   The replication fork and priming of DNA synthesis. The addition of each new nucleo-tide to a free 3'-OH group and the antiparallel directions of the two strands in the double helix allow continuous synthesis on only one of the parental strands (the blue strand). The strand that is synthe-sized continuously is known as leading strand, and its synthesis moves in the same direction as the replication fork. The other strand, the lagging strand, is synthesized in the opposite direction by a discontinuous mechanism: short Okazaki fragments are first synthesized and then ligated together to form an uninterrupted polynucleotide chain. Leading strand synthesis requires only one primer, which is synthesized during replication initiation. Lagging strand synthesis requires multiple prim-ers, each synthesized when a new Okazaki fragment is initiated at the fork. The RNA primers in bacteria are ~10 nucleotides long and are synthesized at the fork by a special enzyme, the primase (a DNA-dependent RNA polymerase, which uses DNA as a template to synthesize the RNA primer).

round-about, but that is the way it is. The small DNA fragments that are inter-mediates in this process were discovered by a Japanese scientist, Reiji Okazaki, in 1968, and are known as **Okazaki fragments**.

It must not be assumed that all of the enzymes involved in DNA replica-tion act independently. Rather, modern research has showed that they are linked together to form an enormous multienzyme complex called a **replisome** which travels as a unit along the DNA. Actually, it is probably more correct to say that the DNA moves through the complex, which is stationary in the cell. In this view, the whole mechanism looks astonishingly like the robot assemblies used in modern factories to carry out multiple operations on product intermediates pass-ing down the assembly line. A view of the *E. coli* replisome is shown in Fig. 8.5.

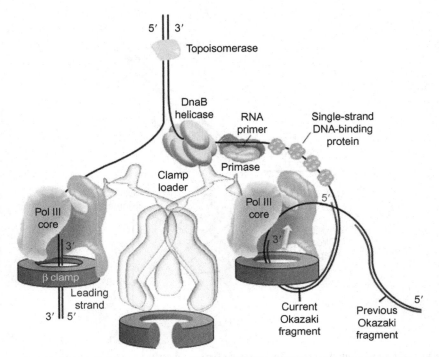

FIG. 8.5 Core proteins at the DNA replication fork. Two DNA core polymerases are active at the fork at any one time, allowing synthesis of the leading and the lagging strands simultaneously. Note the loop (trombone structure) formed by the lagging strand template that allows the two core polymerases to move in the same direction. Both polymerases are anchored to their templates (so that they do not fall off) by auxiliary proteins (a sliding clamp and a clamp loader). The other obligatory protein factors at the fork are:

- DNA helicase, which unwinds the parental helix in an energy-requiring (ATP-dependent) process;
- DNA topoisomerase (not shown), which relieves the superhelical stress accumulating in front of the moving DNA helicase;
- primase (DNA-dependent RNA polymerase), which synthesizes the RNA primers;
- single-strand DNA-binding (SSB) proteins, which cover the single-stranded lagging-strand template to protect it from degradation and hold the DNA in an open conformation with the bases exposed.

(Adapted from Pomerantz, R. T., O'Donnell, M. 2007. Trends Microbiol. 15, 156–164, with permission from Elsevier).

Finally, there are special places on genomes for the initiation and termination of replication. A schematic of the *E. coli* genome is shown in Fig. 8.6. Almost all of the DNA of this bacterium is contained in a circular double-strand molecule of about 4.6 million base pairs. There is one unique site (*ori*) where replication starts; two replisomes are assembled here and proceed around the circle in opposite directions. About half way around the circle are a series of

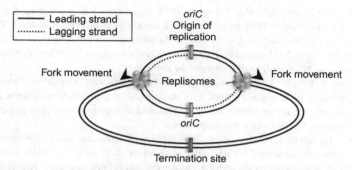

FIG. 8.6   Bidirectional DNA replication of the circular bacterial chromosome. The schematic shows sites for initiation and termination of replication. Replication is initiated at a specific sequence (*oriC*), with two replication forks moving in opposite directions. The orange balls represent replisomes, the protein complexes at the forks that contain the enzyme activities needed for replication to occur. The similarity of the replicating structure to the Greek letter theta (θ) has led to this process being referred to as the "theta mode" of replication.

termination sites. A replication fork encountering one of these can be halted; the fork coming from the opposite direction will likewise be stopped. This leaves two nearly complete intertwined copies of the genome. Any incomplete regions must be repaired, and the copies separated.

We have gone into some detail here to show that what might have appeared to be quite simple, immediately after Watson and Crick published their model of DNA structure, turned out to be exceedingly complex. To unravel the whole picture required decades of work in many laboratories. We continue today in the attempt to understand the exceedingly complex molecular structure that carries out DNA replication. Such research is of much more than academic interest, for the uncontrolled replication of DNA, and hence cell division, is a primary feature of cancer.

EPILOGUE

Understanding how DNA replicates *in vivo*, complex as it is, was only the tip of the iceberg. It was clear by 1953 that at least a major dynamic function of DNA was to provide instructions for the amino acid sequences of thousands of proteins. This became the major unsolved problem in molecular biology, and it immediately attracted the attention of many. But before it could be solved, the problem had to be defined.

FURTHER READING

Book and Reviews

Alberts, B., 2003. DNA replication and recombination. Nature 421, 431–435. A vivid discussion of how the discovery of the DNA structure 50 years back provided new insights into processes like DNA replication and recombination. Argues that new types of approaches involving chemistry, physics, and computational tools are needed to fully understand the complexity of replication.

Cox, L.S. (Ed.), 2009. Molecular Themes in DNA Replication. RSC Publishing, Cambridge, England. An update on DNA replication by scientists actively involved in research. Potential drug targets are also discussed, in the context of treating malaria and cancer.

Hausman, R.E., 2002. To Grasp the Essence of Life. Kluwer Academic Publishers, Dordrecht, The Netherlands. More detail on the history of the research.

Kusic-Tisma, J. (Ed.), 2011. Fundamental Aspects of DNA Replication. Provides new insights into the replication process in eukaryotes, from the assembly of pre-replication complex and features of DNA replication origins, through polymerization mechanisms, to propagation of epigenetic states. In TechOpen (open access book).

Langston, L.D., Indiani, C., O'Donnell, M., 2009. Whither the replisome: emerging perspectives on the dynamic nature of the DNA replication machinery. Cell Cycle 8, 2686–2691. Summarizes results changing views on the replisome as a stable and defined structure. "The replisome is more dynamic than ever thought possible".

Steitz, T.A., 1999. DNA polymerases: structural diversity and common mechanisms. J. Biol. Chem. 274, 17395–17398. Examines the functional and structural similarities and differences among DNA polymerases.

Zlatanova, J., van Holde, K.E., 2016. Molecular Biology. Structure and Dynamics of Genomes and Proteomes. Garland Science, Taylor & Francis Group, New York, NY, pp. 457–482. A detailed description of DNA replication in bacteria. The following chapter in this textbook describes the process in eukaryotes.

Classic Research Papers

Cairns, J., 1963. The bacterial chromosome and its manner of replication as seen by autoradiography. J. Mol. Biol. 6, 208–213. Great informative pictures.

Kornberg, A., 1960. Biological synthesis of deoxyribonucleic acid. Science 131, 1501–1508. The initial announcement of a polymerase. Won him a share of the Nobel Prize.

Okazaki, R., Okazaki, T., Sakabe, K., Sugimoto, K., Kainuma, R., Sugino, A., Iwatsuki, N., 1969. The in vivo mechanism of DNA chain growth. Cold Spring Harb. Symp. Quant. Biol. 33, 129–143. Okazaki fragments.

Chapter 9

The Central Dogma

PROLOGUE

By 1953 both biochemistry and genetics had approached the molecular level. The many decades of developing techniques and collecting data and information had reached the point where the sequences and three-dimensional structures of proteins were about to become accessible (see Chapters 3 and 4). In genetics, DNA had been clearly identified as the material substance of genes, and its 3D structure was revealed by Watson and Crick (see Chapter 7). The coincidence of these massive accomplishments allowed the birth of a true "molecular biology" (Fig. 9.1). But as always, a new paradigm raised new questions. How did DNA replicate to maintain genetic continuity? As we have seen in Chapter 8, that question could be answered, but was more complex than anticipated. How did it utilize the information stored in its sequence to dictate the structures of proteins? How was that information transmitted? These questions were to dominate the next decade, probably the most fruitful in the history of biology.

SPEAKING IN DIFFERENT LANGUAGES

It is often assumed that Watson and Crick's proposal for the structure of DNA immediately opened the doors to understanding the connection between genetics and the everyday workings of the cell. However, even allowing for the notation in the 1953 paper concerning the implications for DNA replication, the picture, even in 1955, was very far from complete. If DNA was indeed the repository and carrier of genetic information for the structure and everyday functions of the cell, how was this information transferred from the DNA to the "working" molecules, the proteins?

It was Francis Crick who first saw this clearly as a "translation" of information from one language to another. A particularly vexing question was how sequence information from DNA sequence, expressed in an alphabet of four different nucleotides, could dictate protein sequences, with an alphabet of about 20 amino acid types. Equally challenging was the "dimension" problem. DNA was clearly a one-dimensional string of information. On the other hand, it was clear by the 1950s that in many, if not most, proteins the one-dimensional polypeptide sequence was folded into a compact three-dimensional structure. That these structures were specific for each protein was suggested by the fact that many

The Evolution of Molecular Biology. https://doi.org/10.1016/B978-0-12-812917-3.00009-7

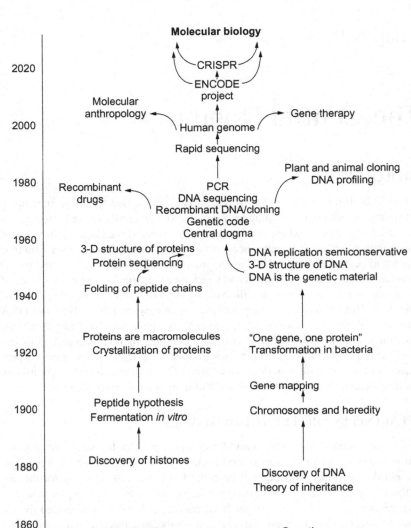

FIG. 9.1 Time-line describing the development of molecular biology from aspects of genetics and biochemistry.

proteins could be crystallized. Indeed, the structures of the oxygen-binding proteins hemoglobin and myoglobin were being studied by crystallography at that moment. Furthermore, it was presumed by most biochemists of the time— including the great Linus Pauling—that the critical feature determining protein function was the folding of their *surfaces*, which was believed to be fine-tuned by interaction with "template" molecules. How could a one-dimensional nucleic acid sequence specify a particular surface, or the structure of the molecule in three dimensions? How much, if any, of the structure was dictated by templates?

Crick simply circumvented this problem by an audacious hypothesis—the one-dimensional sequence of amino acid residues in a protein itself must dictate the three-dimensional folding. This proposal quickly received support from an unexpected source. The American biochemist Christian Anfinsen had been studying the denaturation (unfolding) of the enzyme ribonuclease (which cleaves RNA) by heat or by changes in solvent environment. Much to everyone's surprise, he found the process completely reversible. Enzyme that had been denatured showed no evidence for folded structure, and had lost its activity, but when returned to "native" conditions regained folding and the ability to digest RNA. No "template" was needed. This reduced the DNA/protein problem to a one-dimensional problem. If the DNA sequence is sufficient to dictate the amino acid sequence in a protein molecule, that is all that is needed for the protein to fold itself in three-dimensions and perform its function. Of course, as in all science, the full situation turned out to be more complex, but in 1957 Anfinsen's experiment was sufficient to vindicate Crick's hypothesis.

INTUITING A DOGMA

But *how* did sequence information flow from DNA to protein? The simplest model would seem to be that some kind of complementarity could occur between the surface of the DNA double helix and a polypeptide chain. Almost immediately after publication of the DNA structure, Crick received a letter from George Gamow, a Russian physicist recently fascinated by molecular biology. The letter contained a detailed scheme whereby "pockets" along the double helix, each involving a few nucleotides, could hold specific amino acid residues, allowing them to be connected in the proper order.

In a handwritten note to the **RNA Tie Club** (an informal grouping of young molecular biologists) dated early 1955 Crick systematically demolished the Gamow scheme. He cited sequence limitations that the scheme would impose, which were not observed in the few protein sequences known. Furthermore, there were inconsistencies of spacings in the two kinds of chains, and other awkwardness of the model. Crick proposed instead a radical "**adaptor**" hypothesis, involving a whole class of intermediate molecules that could recognize both DNA sequence and specific amino acids. There was, at the time, not a shred of experimental evidence for the existence of these molecules.

Considering the DNA-to-protein problem led Crick to ponder the more general question: Given the three information-containing types of macromolecules (DNA, RNA, and proteins), between which pairs, and in what directions, was information transfer likely? In a paper published in 1970, Crick details how he thought about the problem. First, he considered all conceivable modes of information transfer between DNA, RNA, and protein, as shown in Fig. 9.2A. Here circular arrows represent *replication*, that is, copying of the sequence information contained in a molecule into new molecules of the same kind. Replication is considered easy and very likely for DNA and RNA, but not for

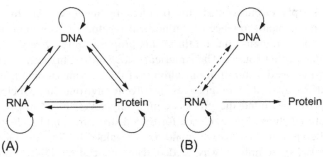

FIG. 9.2 Transfer of information between the three families of macromolecules, DNA, RNA, and protein. (A) All possible directional transfers of sequence information. (B) Directional transfers as proposed by Crick in 1958. *Solid arrows* represent probable transfer, *dotted arrows*, possible transfers. Note in (B) the absence of some of the arrows in (A): there is no transfer of sequence information from proteins, the flow to proteins is unidirectional. *(A, from Crick, F., Nature, 227, 1970, 561–563, Fig. 1; B, modified from Fig. 2.)*

proteins. Of the straight arrows, copying back and forth between DNA and RNA can be easily visualized; it is very similar to replication. Direct transfer between DNA to protein *a la Gamow* is excluded by the arguments given above. The same kinds of arguments make information transfer from protein to DNA or RNA unlikely. This leads to the reduced diagram shown as Fig. 9.2B. All transfers indicated here have in fact been found.

The major conclusion to be drawn from Fig. 9.2B is that genetic information can get from the DNA into protein, but not back out. Proteins do not change genetics, at least directly. This conclusion has come to be known as the "**Central Dogma.**" It was not a dogma but a leap of faith when proposed but has held up remarkably well over the years. We have discovered all of the "allowed" pathways in Fig. 9.2B, and none that were "excluded." It is important to note that the central dogma makes no statements as to how these transfers are made. That was the next big question in 1955.

Transfers back and forth between DNA and RNA were easy to understand because both of these substances "speak the same language." But transfer of sequence information from polynucleotides to proteins was quite another matter. There had to be some kind of code. Furthermore, it could not be a simple 1–1 correspondence between nucleotide and amino acid—that would allow only four amino acids, and it was known that at least 20 different kinds were present in protein sequences. The simplest solution would be a triplet code, with three nucleotides per **codon**. This would allow 64 ($4 \times 4 \times 4$) possible different codons, more than enough to code for all amino acids. All of this was realized by the late 1950s. The Holy Grail now was to discover the code, and how it worked.

WHO IS THE MESSENGER?

However, just cracking the code would not clear up the question as to how genetic information gets from DNA to proteins.

If ideas like the Gamow proposal are wrong, and protein cannot be transcribed directly from DNA, then there had to be some kind of intermediate substance to carry the information from DNA to the protein. This was further demanded by a crucial early observation: in cells of organisms evolutionarily more advanced than bacteria (the **eukaryotes**) DNA seemed always to remain in a separate compartment within the cell, termed the **nucleus**. Yet protein synthesis always took place in the surrounding **cytoplasm**. Thus, there must exist some kind of "messenger" to transmit genetic information from one compartment to the other. It is surprising that in the late 1950s, relatively little attention was paid to this fundamental question. The reason may be that a preconceived, incorrect notion dominated scientists thinking on this problem.

It had long been known that the cytoplasm of all cells was rich in a class of tiny particles called microsomes. Later, when their composition was studied, and found to be rich in RNA, they were renamed **ribosomes** (soma, Greek for body). Ribosomal RNA proved to be the most abundant form of RNA in the cytoplasm, and ribosomes showed evidence to be associated somehow with protein synthesis. Therefore, it was almost natural to assume that ribosomes (and especially their RNA) were the sought-for messages. However, it became clear in the late 1950s that this idea could not be correct. More careful studies of ribosomes showed that these particles, and the RNA molecules they contained, were very homogeneous in size and composition within a cell or even between cells. A collection of messengers, used to code for a variety of proteins of greatly different size and sequence, should not be like that.

More direct evidence came from an experiment conducted by Arthur Pardee, Francois Jacob, and Jacques Monod at the Pasteur Institute in Paris. This has been facetiously named the PaJaMo experiment. They were studying the induction of synthesis of an enzyme upon transfer of DNA from one bacterium to another (Fig. 9.3). They found that enzyme synthesis began as soon as the necessary gene had entered the bacterium, even though no ribosomes were transferred. Therefore, some different entity must be carrying the information from the DNA sequence to direct protein synthesis, and that this began to form as soon as the DNA gene sequence was present. Experiments in other labs showed that destruction of the transferred DNA almost immediately shut down enzyme production. Thus, the messenger must be short-lived, which ribosomes most certainly were not—they had been shown to even survive from one generation of bacteria to the next.

In 1960, Jacob presented these results during an informal discussion with Crick, Sidney Brenner, and a few other molecular biologists during a meeting at Cambridge. According to accounts, the idea of a transient messenger RNA distinct from ribosomal RNA developed in that discussion. The concept was confirmed within 2 years by experiments in a number of laboratories. Most of these depended on infection of E. coli bacteria with bacteriophage T2. Upon infection, synthesis of bacterial proteins quickly ceases, followed by synthesis of new phage proteins. Radioactive labeling of the RNA produced after infection showed that it was transient, associated with "old" ribosomes, and most

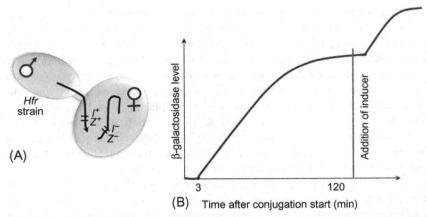

(A)

(B) Time after conjugation start (min)

FIG. 9.3 The PaJaMo experiment. (A) The principle: a *Hfr* high-frequency-recombination (male) strain of bacteria carrying the β-galactosidase gene (*lacZ*) and the repressor gene (*lacI*) was allowed to conjugate with an excess of females lacking these genes. During conjugation, part of the male chromosome containing the wild-type genes was transferred to the female cell, allowing the production of β-galactosidase, encoded by the Z gene. (B) Time-course of β-galactosidase generation. Aliquots of the conjugating bacteria were agitated in a Waring blender to stop conjugation at different times, and β-galactosidase activity was assayed. After a brief lag period (~3 min), the activity rises because more and more female cells have the *lacZ* gene. However, the repressor gene is also being taken up during conjugation, so enzyme production slows and finally halts when repressor has accumulated enough to shut down the expression of the *lacZ* gene. After this time point, the enzyme becomes inducible because the addition of inducer will cause release of this repressor from the DNA.

importantly, that it was complementary in sequence to one strand of the phage DNA. All of these experiments supported the idea of **messenger RNA**, with the ribosomes acting as "staging areas" where the translation of messenger RNA information into protein sequences could occur. These insights divide the transfer of information from DNA to protein into two distinct stages; **transcription** of messenger RNA from a DNA template, followed by **translation** of that message into a protein sequence, while bound to ribosomes. With elaborations, this is the picture held to the present day (Fig. 9.4).

THE GREAT DECADE: 1952–62

The brief period from 1952 to 1962 must be considered one of the most remarkable times in the history of biology. Consider what we knew in 1952. DNA was suspected to be the genetic material, but not everyone was convinced; the Hershey-Chase experiment, which really settled the question was only published in 1952. DNA structure was unknown; in a year, it would be elucidated; and the structure would point to a mode of replication, which would be confirmed by Meselson and Stahl within the decade. The Central Dogma concerning how

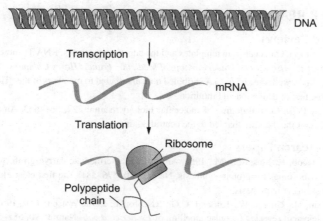

FIG. 9.4   Transcription and translation: the two phases of informational transfer between protein-coding genes and their encoded protein product. The RNA strand is complementary to one of the DNA strands; enzymes called RNA polymerases move along the DNA to copy the information into a messenger RNA molecule. The messenger RNA is then attached to ribosomes, which connect amino acids in the order dictated by the nucleotide sequence in the messenger and the genetic code.

information could be transferred from DNA to protein was proposed by Crick, and "messenger" RNA was both hypothesized and demonstrated in this decade. The role of the ribosome was finally clarified. The necessity for a "code" to translate DNA information into protein information was recognized, and first steps in breaking the code had been made.

In the realm of proteins, advances were no less spectacular. In this same decade, the first protein sequences were determined, and X-ray diffraction provided the first detailed information on the three-dimensional structures of protein molecules. Experimental studies showed that the folding of proteins to form these complex structures could be dictated by protein amino acid sequence. Thus, a direct connection between genetics and biochemistry could be visualized.

EPILOGUE

In short, a whole new area of biology, "molecular biology," had been born from genetics and biochemistry between 1952 and 1962. There was much yet to do, but the questions were now clear. There must be some kind of **genetic code** governing the translation of DNA sequence into protein sequences. But what kind of code, and how did translation occur? How was this controlled in different cells and tissues? As is often the case in science, major advances only occur when the right questions can be asked. Breathtaking surprises lay ahead, to be described in the next several chapters.

FURTHER READING

Books and Reviews

Crick, F.H.C., 1955. On degenerate templates and the adaptor hypothesis. RNA Tie note. (*Welcome Library for the History and Understanding of Medicine. Francis Harry Compton Crick Papers* http://archives.wellcome.ac.uk/.*) A scribbled note distributed to members of the "Tie Club." An incredible feat of deduction and intuition.

Crick, F.H.C., 1970. Central dogma of molecular biology. Nature 227, 561–563. An analysis, in retrospect, of the thinking that led to the central dogma.

Classic Research Papers

Brenner, S., Jacob, F., Meselson, M., 1961. An unstable intermediate carrying information from genes to ribosomes for protein synthesis. Nature 190, 576–581. The first clear elucidation of the "messenger" hypothesis.

Gros, F., Hiatt, H., Gilbert, W., Kurland, C.G., Risebrough, R.W., Watson, J.D., 1961. Unstable ribonucleic acid revealed by pulse labelling of *Escherichia coli*. Nature 190, 581–585. One of several experimental demonstrations of messenger RNA.

Hall, B.D., Spiegelman, S., 1961. Sequence complementarity of T2 DNA and T2 specific RNA. Proc. Natl. Acad. Sci. USA 47, 137–146. The title says it all.

Pardee, A.B., Jacob, F., Monod, J., 1959. The genetic control and cytoplasmic expression of "inducibility" in the synthesis of galactosidase in *E. coli*. J. Mol. Biol. 1, 165–169. The "PaJaMo" experiment.

Chapter 10

The Genetic Code

PROLOGUE

By about 1955, it was evident to the newly born field of molecular biology that genetic information, in the form of DNA sequences, was somehow expressed in protein sequences. It was even becoming clear that RNA played some role as an intermediary in this transfer. But because DNA and RNA store information in sequences of four kinds of nucleotides, and proteins are expressed in a language of 20 kinds of amino acids, there must exist some kind of **code** to relate one to the other. But what *kind* of code, and what were the rules to decipher it by? For the next decade, this would become the compelling interest of a host of molecular biologists.

HOW MIGHT A CODE FUNCTION?

In the mid-1950s, little was understood about the synthesis of proteins. Much of what there was come from years of painstaking work by Paul Zamecnik and his group, who had been using radiolabeled amino acids to attempt to define the constituents necessary for protein synthesis in the test tube. An energy source like adenosine triphosphate (ATP) was found to be required, in addition to microsomes (now called ribosomes) and RNA. It was realized after some time that two distinct kinds of RNA were needed—exogenous RNA added to the mix and a fraction of slowly sedimenting "soluble" RNA from the cell extract itself. With these, a very small amount of radiolabel could be incorporated into material that identified as protein-like. Toward the end of the decade, the significance of these results was becoming clear in the light of new knowledge. The exogenous RNA might correspond to the "messenger" that was being talked about, the "soluble" RNA could have some connection to Crick's "adaptors," and the ribosome now began to look like the site of protein synthesis. These concepts were essential for the search for codons.

WHAT KIND OF CODE?

The idea of a code was by no means new. Indeed, in 1943 Schrödinger had written in "What is Life" about a "code-script" which he supposed to be colinear with the chromosome and contained all the information as to what an organism

The Evolution of Molecular Biology. https://doi.org/10.1016/B978-0-12-812917-3.00010-3

could be. However, when scientists began asking what a code that could translate from polynucleotides to polypeptides might be like, it became evident that there were a bewildering number of possibilities. As noted in Chapter 9, it was felt by some that the codons might exist as sites on the surface of the DNA itself as proposed by Gamow; alternatively, they might utilize the hypothetical "adaptor" molecules proposed by Crick. There was no information as to number of letters (bases) to code for one amino acid, provided at least 20 kinds of amino acids could be coded for. Codons might "overlap" or not (Fig. 10.1A). With overlap, one and the same base in the DNA sequence would be used to code for three consecutive amino acids in the protein sequence, serving, at the same time as the last base letter for one amino acid and as the first or second base letter for the following amino acid. In addition, there could be some base that

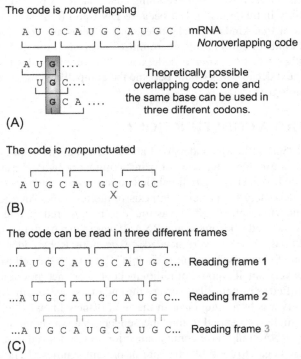

(A)

The code is *non*overlapping

A U G C A U G C A U G C   mRNA
                          *Non*overlapping code

A U G ....
  U G C....   Theoretically possible
    G C A ....   overlapping code: one and
                       the same base can be used in
                       three different codons.

(B)

The code is *non*punctuated

A U G C A U G C U G C
              X

(C)

The code can be read in three different frames

...A U G C A U G C A U G C...   Reading frame 1

...A U G C A U G C A U G C...   Reading frame 2

...A U G C A U G C A U G C...   Reading frame 3

FIG. 10.1 Basic features of the genetic code. (A) Schematic of the two possible ways a triplet code can be read: in the overlapping code; the same base can be part of three codons, whereas in the nonoverlapping code each base is part of only one codon. The genetic code used in nature is nonoverlapping. (B) The code is nonpunctuated: all nucleotides are read in succession and no gaps are allowed between successive codons. (C) The concepts of reading frame. Each nucleotide sequence is read starting from a particular nucleotide. The specification of this first nucleotide defines the succession of codons, and thus, the primary structure of the polypeptide chain. If a different first nucleotide is specified, the same sequence can be read in a different frame, producing a different polypeptide.

acted as a punctuation mark between codons, or the code could be unpunctuated (Fig. 10.1B). Each of these alternatives placed certain constraints on how genes could be expressed. The many possibilities and permutations to the problem attracted cryptanalysts and mathematicians. Debate was extensive, and mostly unprofitable; we shall not bore you with details. Instead, we consider certain crucial experiments that resolved the question unambiguously.

The first involved one of the few forays into lab-bench genetics ever taken by Francis Crick. It had been noted that mutants of the bacteriophage T4 produced by proflavin showed a curious behavior; the induced change in phenotype caused by the mutation could sometimes be reversed by a *second* mutation elsewhere. Proflavin is a flat molecule about the size of a purine base, and it can intercalate between base pairs, causing base-pair deletions or insertions (Fig. 10.2A). After examining many such mutants, Crick and Sidney Brenner concluded that there were of two types: in one case, (+), the proflavin inserted in the template chain in a region not yet copied and produced a **frameshift**, by forcing the addition of another nucleotide in the new chain. In the opposite case, (−), the insertion was in the copied chain and resulted in the subsequent loss of a nucleotide, with an opposite frameshift (Fig. 10.2B). Obviously, this explained that reversion of phenotype could sometimes be accomplished by a second, compensating mutation. More important, it showed that the code must operate on defined reading frames in the cell. A further observation, that three mutations of the same sign could produce compensation, was the first evidence that the codon size was three. These results began the transition from idle speculation to experimental study of the code.

WHAT WERE THE CODE WORDS?

If codons were three nucleotides in length, the four DNA nucleotides could potentially code for 64 amino acids; however, there are only about 20 amino acids, which means that if all codons are used, there should be redundancy (degeneracy) in the code; in other words, one and the same amino acid could be

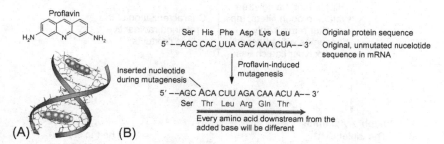

FIG. 10.2   The experiments that defined the triplet nature of the genetic code. (A) The structure of the intercalator proflavin and structure of the DNA double helix with intercalated proflavin. (B) A schematic illustrating how the experiment worked.

coded by more than one codon. To deduce even one codon corresponding to a particular amino acid seemed a formidable task. But the first success came not to one of the giants in the field, but to two unknown, modest young researchers at the National Institutes of Health, in Bethesda, Maryland. There, Marshall Nirenberg and his assistant Heinrich Matthaei were attempting to improve the efficiency of in vitro protein synthesis, using the bacterial cell extracts that had been developed by Paul Zamecnik over several years. Radioactively labeled amino acids were added to these extracts, along with RNA to serve as a template for the assembly of a polypeptide chain, and radiolabeled protein was sought (Fig. 10.3). Results had been discouraging for years. On May 22, 1961, while Nirenberg was out of town, Matthaei did one more of many experiments. But instead of natural RNA, he used a synthetic product, polyU, a long-chain polymer of U residues. To their astonishment, Nirenberg and Matthaei found much enhanced incorporation of amino acids into product, and this was virtually all in

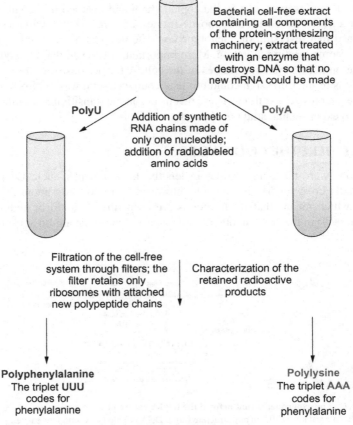

FIG. 10.3   The cell-free experiments of Nirenberg and Matthaei.

polyphenylalanine. Thus, they had discovered the first code word: UUU (assuming a triplet) coded for the amino acid phenylalanine!

In the summer of 1961, Nirenberg attended the International Congress of Biochemistry in Moscow. Because he was essentially unknown in the field, his talk was assigned to a minor category and was attended by few people and even fewer of the "club." But Mathew Meselson was there and immediately realized the remarkable importance of the work. He persuaded Crick, who was in charge of such matters, to allow Nirenberg to speak again. This time, a much larger hall was packed, and Nirenberg and Matthaei were instantly famous. Their fame was not met with joy by all members of the establishment. How could these unknowns from America have cracked the problem that had stumped us for almost a decade?

The chase for more codons began immediately, starting with homopolymers like ...AAAAA... which coded for polylysine (Fig. 10.3). But these could provide only a few codon assignments. More were obtained by using polymers containing dimer or trimer repeats. Production and use of such defined polynucleotides was pioneered by a skilled Indian-American biochemist, Gobind Khorana (Fig. 10.4). By combining results like this, more codons could be deduced. In years to follow, as protein synthesis was better understood, very simple ways of making codon assignments were developed. The last pieces were put together in the late 1960s.

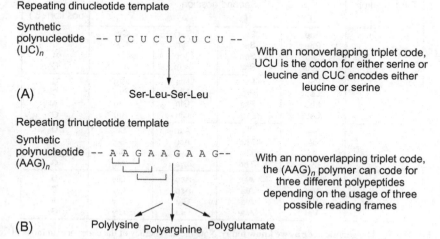

Repeating dinucleotide template

Synthetic
polynucleotide   -- U C U C U C U C U --
$(UC)_n$

With an nonoverlapping triplet code,
UCU is the codon for either serine or
leucine and CUC encodes either
leucine or serine

(A)   Ser-Leu-Ser-Leu

Repeating trinucleotide template

Synthetic
polynucleotide   -- A A G A A G A A G--
$(AAG)_n$

With an nonoverlapping triplet code,
the $(AAG)_n$ polymer can code for
three different polypeptides
depending on the usage of three
possible reading frames

(B)   Polylysine Polyarginine Polyglutamate

FIG. 10.4   Deciphering the genetic code by using repeating dinucleotide or trinucleotide synthetic polymers as templates for protein synthesis. These types of experiments indicated possible codons for the amino acids that became incorporated into the polypeptide chains, but the final codon assignment required further experiments. The most definitive answers came from using simple synthetic trinucleotides as templates and identifying the nature of the tRNA molecules that became attached to the ribosomes, in the first step of protein synthesis.

THE CODE

The code is usually expressed in the form of a Table like that shown as Fig. 10.5. Each of the codons is assumed to be written in the 5′ to 3′ direction, so the "First Position" is at the 5′ end. Examination of the table reveals several features of the code.

- The code is **degenerate**, with several codons for each amino acid (except tryptophan).
- All triplet codons are used, but three (UAG, UGA, and UAA) serve as "stop" signals to end translation of a messenger RNA and release the new polypeptide chain from the ribosome. Codon AUG codes for methionine, but in certain situations will also serve as a "start" signal; in other words, it will determine the exact position in the mRNA where the polypeptide chain should begin.
- The degeneracy of the code is mainly expressed in the third position. Crick proposed an explanation for this in terms of possible "wobble" of base-pairing in this position (Fig. 10.6).
- This code is almost, but not quite, universal. With a very few examples of unusual codon assignment, it is the same in every creature, living or fossil that has been examined. This speaks strongly for their common origin.

First position (5′)	Second position				Third position (3′)
	U	**C**	**A**	**G**	
U	Phe	Ser	Tyr	Cys	U
	Phe	Ser	Tyr	Cys	C
	Leu	Ser	Stop	Stop	A
	Leu	Ser	Stop	Trp	G
C	Leu	Pro	His	Arg	U
	Leu	Pro	His	Arg	C
	Leu	Pro	Gln	Arg	A
	Leu	Pro	Gln	Arg	G
A	Ile	Thr	Asn	Ser	U
	Ile	Thr	Asn	Ser	C
	Ile	Thr	Lys	Arg	A
	Met	Thr	Lys	Arg	G
G	Val	Ala	Asp	Gly	U
	Val	Ala	Asp	Gly	C
	Val	Ala	Glu	Gly	A
	Val	Ala	Glu	Gly	G

FIG. 10.5   The genetic code as we know today. Note the degeneracy of the code: one amino acid may be specified by multiple codons (e.g., serine has 6 codons, glycine has 4, etc.). The first two nucleotides are often enough to specify a given amino acid (e.g., serine is specified by UC). Note also that codons with similar sequences specify amino acids of similar chemical properties (e.g., serine and threonine differ in first letter; aspartic acid and glutamic acid in the third letter). This ensures that many mutations will result in the incorporation of a similar amino acid that would not significantly affect the structure/function of the protein.

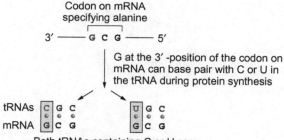

Codon on mRNA
specifying alanine

3' ——— G C G ——— 5'

G at the 3'-position of the codon on
mRNA can base pair with C or U in
the tRNA during protein synthesis

tRNAs   C G C   U G C
mRNA   G C G   G C G

Both tRNAs containing C or U carry
the same amino acid, alanine.

FIG. 10.6   The Wobble hypothesis as suggested by Crick. The first two bases in the mRNA codon form hydrogen bonding with their corresponding bases on the tRNA anticodon in the usual Watson-Crick manner, in that they only form base pairs with complementary bases. However, third base in a codon in the mRNA can undergo non-Watson-Crick base pairing with the first base of the respective base triplet (anticodon) on the tRNA. Thus, one tRNA molecule (with one amino acid attached) can recognize and bind to more than one codon, due to the less-precise base pairs at this position. The hypothesis explains why multiple codons can code for a single amino acid.

EPILOGUE

Learning the code was essential to understanding of the transmission of genetic information from DNA to protein. But meanwhile there was clearly much more to do to complete the picture. Acceptance of the idea of "messenger" RNA in the 1960s meant that there was another process that must be understood—**transcription** of the proper DNA sequence into RNA. That messenger RNA must then be **translated** to produce a polypeptide, the protein product. Understanding these processes and their regulation is a massive task that is still underway. The next chapter describes the pioneering work in this area.

FURTHER READING

Books and Reviews

Crick, F.H.C., 1966. The genetic code – yesterday, today, and tomorrow. Cold Spring Harb. Symp. Quant. Biol. 31, 3–9. A review near the end of the chase.

Zamecnik, P.C., 1979. Historical aspects of protein synthesis. Ann. N. Y. Acad. Sci. 325, 269–301. From one of the pioneers in the field.

Classic Research Papers

Crick, F.H.C., 1966. Codon-anticodon pairing; the wobble hypothesis. J. Mol. Biol. 19, 548–955. Another major contribution by Francis Crick.

Crick, F.H.C., Barnett, L., Brenner, S., Watts-Tobin, R.J., 1961. General nature of the genetic code for proteins. Nature 192, 1227–1232. Description of the T4 mutation experiments.

Nirenberg, M.W., Matthaei, J.H., 1961. The dependence of cell-free protein synthesis in E. coli upon naturally occurring or synthetic polyribonucleotides. Proc. Natl. Acad. Sci. USA 47, 1588–1602. This paper describes the polyU experiment, in a modest way.

Chapter 11

Gene to Protein: The Whole Path

PROLOGUE

Despite the enormous advances achieved in the 1950s, the problem posed by the central dogma was still unresolved: just *how* did a DNA sequence transmit its information to specify a protein sequence? Even by 1960, the whole pathway was still uncharted. There were, to be sure, bits and fragments of the map lying about, but there seemed to be no clear way to put them together. Then, in a very short period, all coalesced into a coherent picture.

WHAT WAS KNOWN IN 1960?

To present a coherent time-line for progress in the protein synthesis problem between 1950 and 1960 seems impossible. There was just too much happening simultaneously, in divergent directions. So, we shall follow advances in several different aspects and then describe how these came together to produce a new view.

Messenger RNA. By 1960, the idea was prevalent that there must be some kind of an intermediate, probably RNA, between DNA and protein. There was in fact evidence, as early as 1955. Elliot Volkin and Lazarus Astrachan, at the Oak Ridge National Laboratory in Tennessee, labeled *Escherichia coli* that were infected with T2 phage with radioactive phosphorus and observed a remarkable result: the labeled RNA resembled in base composition the T2 DNA, not the bacterial DNA. They boldly suggested, after repeating the experiment with other phage, that the RNA might be coding for phage proteins. This is exactly right, but the work was wholly neglected and little referred to—except that Crick, in 1960, admitted to being puzzled by it. Strong support for this evidence for a messenger came from work by a young scientist, Benjamin Hall, in Sol Spiegelman's lab at the University of Illinois. He repeated the Volkin-Astrachan experiments but added the powerful method of DNA-RNA **hybridization**. In these experiments, the DNA is heated to separate the strands of the double helix, mixed with an excess of the RNA to be tested, and slowly cooled. If there is complementarity between the RNA and DNA, this can be detected by the formation of hybrid molecules containing both polynucleotides (Fig. 11.1). It was observed that the RNA hybridized to the phage DNA, not the bacterial DNA,

The Evolution of Molecular Biology. https://doi.org/10.1016/B978-0-12-812917-3.00011-5

DNA fragments

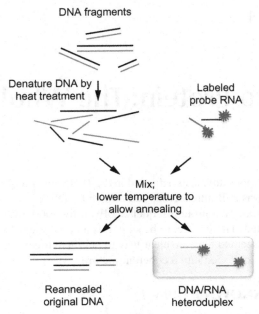

Denature DNA by
heat treatment

Labeled
probe RNA

Mix;
lower temperature to
allow annealing

Reannealed
original DNA

DNA/RNA
heteroduplex

FIG. 11.1 Nucleic acid hybridization. The reaction can be performed in solution or on filter paper; in the latter case, the DNA fragments are first fractionated by gel electrophoresis, transferred to filter paper where the interaction with the labeled RNA probes is allowed to occur.

strongly supporting the Volkin-Astrachan conclusion. The famous PaJaMo experiment (Chapter 9) also pointed to an ephemeral message, although this seems not to have been realized at first.

Ribosomes. If there was a messenger RNA in the cell, where was it? It was expected to be found in the cytoplasm, for in higher organisms, protein synthesis was known to occur there, not in the nucleus. Now, there was an enormous pool of RNA in the cytoplasm of all cells that had no known function. This was in **ribosomes**, originally called microsomes. They were observed by Albert Claude as early as 1940 using darkfield microscopy, where they appeared in cytoplasmic extracts as hordes of tiny particles. A decade later their study was taken up by the Rumanian-American biologist, George Palade. His electron microscope studies showed them to be immense particles on the molecular scale, about 10 nm in diameter. Microsomes are now called ribosomes because they were found to contain about equal fractions of RNA and protein (Fig. 11.2). What function could all that RNA serve except as messenger? This concept was probably held by a majority of molecular biologists in the 1950s, but there were obvious problems with it. Proteins came in a wide variety of sizes, so one might expect ribosomes to mimic this. But ribosomes, and even their RNA, were surprisingly uniform in size and composition within a cell, or between cells, or even between organisms.

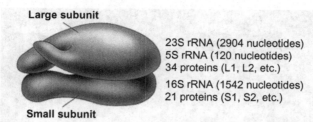

FIG. 11.2   The ribosome: an assembly of one large and one small RNP subunit, each containing rRNA molecules and a set of bound ribosomal proteins. The model is based on low-resolution EM images of the 70S ribosome. The two subunits are usually referred to by their sedimentation coefficients (S values).

Then, there was the question of durability. Several kinds of experiments, such as those detailed above, and the PaJaMo experiment (Chapter 9) indicated that mRNA was very short lived, but ribosomes were known to be remarkably stable, even surviving through bacterial cell division. Despite all of this, the "ribosome-gene" theory survived until 1960, although very little was known about the ribosomes themselves. The picture available in 1960 resembled Fig. 11.2, it was known that the particle consisted of two subunits, and contained both RNA and protein.

Over the decades, numerous laboratories have been involved in efforts to understand ribosome structure in detail. The composition of the particles in terms of the types of RNA and proteins (Fig. 11.2) has been determined for ribosomes from both bacterial and eukaryotic cells. In addition, the morphology of ribosomes has been studied by ever more sophisticated imaging techniques, including traditional microscopy, electron and cryo-electron microscopy, and X-ray crystallography. The current resolution of ribosome structure derived from such methods is amazing (Fig. 11.3). It came as no surprise that such results attracted the attention of the Nobel prize committee, which awarded the 2009 Prize in Chemistry to Venkatraman Ramakrishnan, Thomas A. Steitz, and Ada E. Yonath for their "studies of the structure and function of the ribosome."

BREAKTHROUGH

It can be argued that all of the information needed to resolve the puzzle of protein synthesis was available before 1960. In any event, the resolution of all of these uncertainties can be dated. According to reliable accounts, it occurred on Good Friday, 1960, in Sydney Brenner's rooms at Cambridge University, where a small group had gathered to discuss recent research in molecular biology. Most had been at a scientific meeting in London. Present, in addition to Brenner, were Francis Crick, Francois Jacob, Leslie Orgel, Alan Garen and wife Susan, and perhaps Ole Maaloe. Discussion began with a description by Jacob of recent work in Paris to refine and further substantiate the PaJaMo

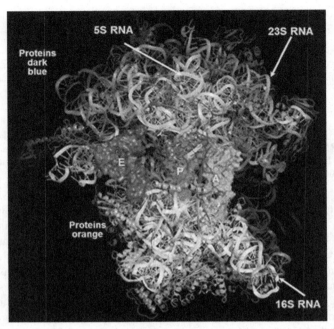

FIG. 11.3   Structural models of the bacterial ribosome derived from X-ray crystallography. The resolution of this structure is 2.8Å. The intact 70S ribosome is complexed with mRNA and tRNA. Nestled between the two subunits are the three sites for binding of tRNAs: A(aminoacyl)-site, P(peptidyl)-site, and E(exit)-site. The A site is the site of binding of the incoming tRNA carrying its respective amino acid; the P site interacts with tRNA that carries the nascent polypeptide chain; the E site interacts with the deacylated tRNA; in other words, the tRNA that has already given its amino acid to the growing chain. The mRNA at these three sites is just barely visible. *(From Ramakrishnan, V., 2008. Biochem. Soc. Trans. 36, 567–574, Fig. 5, with permission.)*

experiment (see Chapter 9). Talk continued about other experiments relative to the messenger, including the "old" work of Volkin and Astrachan, and the recent unpublished studies by Hall and Spiegelman, about which Brenner had seen preliminary information.

"That's when the penny dropped and we realized what it was all about" Crick has been quoted. And—of course—the messenger RNA was an ephemeral, short-lived substance that was synthesized complementary to one DNA strand; it attached to the ribosome where it recruited the amino acids to be added to a growing polypeptide chain. The ribosome was just a molecular machine that served as a site for the synthesis to occur and somehow aided in the process. (Until very recently, the ribosome was considered to be the actual catalyst of the chemical reaction of amino acid addition, but now this role has been assigned to a portion of the tRNA carrying the amino acid to be attached.) The same ribosome could work with any message—that is why the ribosomes were all the same! In just a few moments the confusion of decades was cleared away. There were still many aspects to clarify—the work continues to the present day—but a major intellectual log jam had been broken.

There was a flurry of publication. Brenner, Jacob, and Matthew Meselson collaborated at Caltech to demonstrate, via density gradient centrifugation, that new RNA generated after phage infection bound to old ribosomes. Alexander Rich, at MIT, directly demonstrated DNA-RNA hybridization by mixing polyadenylate (for RNA) with deoxy polyuridylate (for DNA). Jacob and Monod wrote a long review which summarized recent developments in the field, including the concept of mRNA. Surprisingly, no other participants at the crucial Cambridge meeting were included as coauthors.

THE REST OF THE STORY

There still existed major gaps in the picture. How was the mRNA synthesized as a copy of a DNA strand? What was the connection between RNA codons and amino acid residues established on the ribosome?

RNA Polymerase. The existence of mRNA molecules corresponding to particular proteins necessitated a mechanism to make complementary copies of specific regions of one strand of duplex DNA in the genome. This resolves into two problems: the copying process itself and its control.

Practically all biochemical processes are catalyzed by specific enzymes, so it was not surprising that a search for an enzyme to facilitate the polymerization of ribonucleotides on a DNA template occurred, just as Arthur Kornberg was looking for a DNA polymerase (see Chapter 8). Indeed, in 1960 Gerald Hurwitz reported such activity in cell extracts. Importantly, it required all four nucleotide triphosphates, as well as a template DNA. In another year, the enzyme was purified by Michael Chamberlin, and shown to be a large molecule, comprised of four subunits. But something was missing, for while this preparation would catalyze formation of RNA, it was indiscriminate as to start and stop positions, or even which strand to copy. It took until 1969 and more gentle methods of preparation to discover that there was a more-weakly bound fifth subunit. When this was present, polymerization began only at specific **promoter** sites on the DNA and proceeded to defined termination. A visualization of our present knowledge of the structure of the bacterial RNA polymerase enzyme in action is shown in Fig. 11.4. Note how the enzyme opens a "bubble" by locally unwinding the DNA. This allows one strand to be available for copying.

Discovery of promoter sites on DNA provided the first insights into how transcription could be controlled. This has become a vast field, with implications in many areas of biology and medicine. The first studies in this area to produce real insights were carried out with bacteria and viruses. We describe such experiments in the next section of this chapter.

Transfer RNA (tRNA). By the late 1950s, it was clear to almost everyone in the field that protein sequence could not be read directly off DNA: something

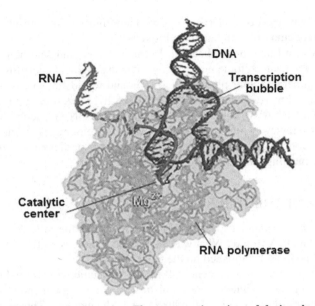

FIG. 11.4 RNAP from the bacterium *Thermus aquaticus* pictured during elongation. Note the transcription bubble in the DNA: melting of a stretch of the double helix exposes bases to be used as templates for addition of incoming RNA precursors, ribonucleotides. Portions of the enzyme were made transparent so as to make the path of RNA and DNA more clear. The magnesium ion is located at the enzyme active site. *(From Wikipedia.)*

like Crick's adaptor was needed. In 1958 Mahlon Hoagland, working in Paul Zamecnik's laboratory, made a most peculiar (and accidental) observation. He was studying the incorporation of radiolabeled ATP into RNA in cell-free liver extracts. As a control, to show that the washing procedure to remove unincorporated ATP was sufficient, Hoagland did a "dummy" experiment in which he added radiolabeled leucine instead of ATP. An amino acid like leucine should not be found in RNA. Yet it was abundantly! Further analysis showed that the leucine became attached to small RNA molecules found in the supernatant of cell extracts. These could be the adaptors! Further research showed that there were a host of these kinds of molecules, at least one specific for each kind of amino acid. The amino acid was bound as an aminoadenylate, an activated form produced as shown in Fig. 11.5. A specific amino acid would be so attached only to a tRNA with anticodon corresponding to that amino acid according to the genetic code. There was a specific enzyme to form each tRNA-amino acid pair. This is how the code works. The function of these adaptor molecules was now recognized, and they were titled **transfer RNAs** (tRNAs).

In 2 years Zamecnik reported partial purification of tRNA, and by 1965 Robert Holley had not only purified a specific alanine tRNA but had determined

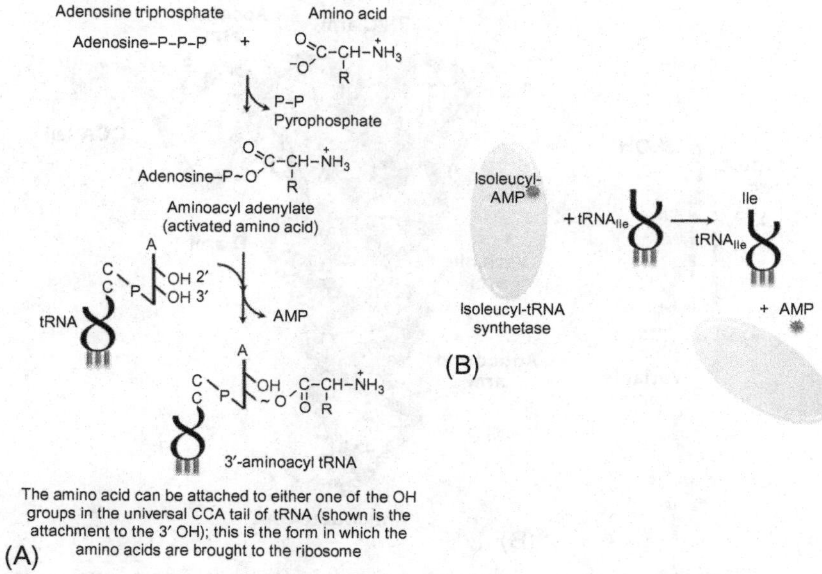

Adenosine triphosphate   Amino acid

The amino acid can be attached to either one of the OH
groups in the universal CCA tail of tRNA (shown is the
attachment to the 3′ OH); this is the form in which the
amino acids are brought to the ribosome

(A)

(B)

FIG. 11.5   Aminoacylation of tRNA: the chemistry. (A) Amino acid activation through attachment of adenosine monophosphate. The pyrophosphate that is released during the first step is immediately hydrolyzed by an abundant enzyme pyrophosphatase to two molecules of inorganic phosphate. This prevents the amino-acid activation reaction from going into reverse. At the next step, an ester linkage between the amino acid and the ribose of the terminal A in the universal CCA tail of the tRNA is created. This is the form in which the amino acids are brought to the ribosome for incorporation into the nascent polypeptide chain. Both steps occur on the aminoacyl-tRNA synthetase, as illustrated in panel (B).

its sequence. Holley shared the 1968 Nobel Prize in Physiology or Medicine with Gobind Khorana and Marshall Nirenberg for "their interpretation of the genetic code and its function in protein synthesis."

As shown in Fig. 11.6A, the sequence immediately suggested hydrogen bonding to produce a specific three-dimensional (3D) structure. That structure has been determined at high resolution (Fig. 11.6B) and is remarkably informative; the acylated amino acid binds to the 3′-end of the chain, whereas an exposed loop (bottom) carries the anticodon, which will bind to the codon on the mRNA, when both tRNA and mRNA are bound to the ribosome. As the mRNA passes through the ribosome, each codon occupies a specific place, to accept a tRNA with its specific amino acid. That amino acid can then be added to the polypeptide chain. We now know an enormous amount about the ribosome and its mechanisms, as shown in Fig. 11.3.

The ribosome also responds to signals on the mRNA for initiation and termination of translation. As we mentioned in Chapter 10, a methionine tRNA, carrying a modified methionine, and the anticodon to the methionine codon, AUG,

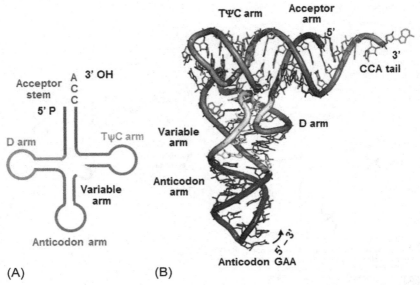

FIG. 11.6 Primary, secondary, and tertiary structure of transfer RNA (tRNA). (A) Generalized cloverleaf secondary structure of any tRNA. (B) The crystal structure reveals the 3D organization of the helices. Note the exposed bases that form the anticodon; this organization helps the interaction with the codon on the mRNA. Some arms derive their names from specifically modified residues that occur frequently in these parts of the polynucleotide: *D*, dihydrouridine; *T*, thymine ribonucleotide; ψ, pseudouridylate. The amino acid attaches through its COOH group to the 3′- or the 2′-OH groups of the ribose of the terminal adenine (A) residue in the universal CCA triplet at the acceptor stem's terminus. The anticodon arm contains the triplet anticodon which specifies the amino acid that is to be incorporated into the growing peptide chain by interacting with the codon of the mRNA. *(Adapted from Wikipedia.)*

are essential. This will attach to a free small subunit of the ribosome in the proper place to recruit the large subunit, the messenger RNA, and begin translation. Translation will continue until one of the "stop" codons (UAA, UAG, or UGA) is encountered, at which point the polypeptide chain will be released, and the ribosome will dissociate into its small and large subunits, ready to begin another round. This description is greatly oversimplified; there are a number of protein factors associated with initiation and termination.

REGULATION OF TRANSCRIPTION IN BACTERIA

Bacteria must respond to changes in their environment which often involve changes in nutrient. The classic "model bacterium" *E coli* is very adept at this. To take one example, which has played a major role in molecular biology, we consider the metabolism of milk sugar, lactose. Lactose is a **disaccharide** consisting of two simple sugars, glucose and galactose, linked together. To utilize lactose as a nutrient, lactose must be cleaved into those products. *E. coli* has an enzyme, β-galactosidase, that catalyzes this cleavage. In 1940, a young French scientist at the Sorbonne, Jacques Monod, made the peculiar discovery that *E. coli* fed a mix of glucose and lactose used up the glucose first before beginning consumption of lactose. World War II interrupted the work; Monod served in the French army and then in the underground resistance movement against the Nazis. To avoid detection, he moved to the Institute Pasteur to continue his research.

In 1954, Monod was joined by another French scientist, Francois Jacob. They found that synthesis of β-galactosidase could be stimulated by lactose, or a number of related compounds. These they called **inducers**. Surprisingly, the production of some related proteins was simultaneously enhanced; these proteins include a **permease**, which helps to get lactose into the cells and a transacetylase enzyme that detoxifies imperfect products of the pathway. Remarkably, they found mutants occurring *outside* of the three genes which could influence the trio. To explain these observations, they postulated that the three genes must all be under the control of a fourth gene (the *Lac I* gene), which coded for a fourth protein, which was named the **repressor**. The gene map, which is now known to be correct, is depicted in Fig. 11.7. All four genes are considered parts of one **operon**, a module of genes needed for the utilization of lactose by *E. coli*. It functions as follows: in the absence of lactose, repressor is bound to the **operator** site, blocking the promoter from beginning transcription of the operon. When lactose or any other inducer is present in sufficient amount, it displaces repressor from the operator, and transcription of the operon can commence. This is gene activation by relief of repression. It is also often found that regulation of operons can occur by direct activation of operators (see the role of CRP activation of the *lac* genes in Fig. 11.7). There are many, many possible metabolic pathways in a cell, and enormously complex networks of regulation have evolved to make sure that the right pathways function (or do not function) at appropriate times.

This work had another important spinoff. The observation that the binding of one substance (the inducer) to a protein could influence that protein's affinity for another substance (the operator site on DNA) led Jacob and Monod to advance, in 1963, the idea of **allosteric** interactions. In such interactions, the binding of one molecule to a macromolecule modifies the macromolecules affinity for binding another. The control in the *lac* system is a good example: binding of the inducer to the repressor causes the latter to be released from its interaction with DNA. This kind of regulation has applications in many areas of biochemistry. It often takes the form of **feedback regulation**: the product of a long chain of reactions may inhibit the first enzyme in the chain. Then, energy is not wasted in making something no longer needed.

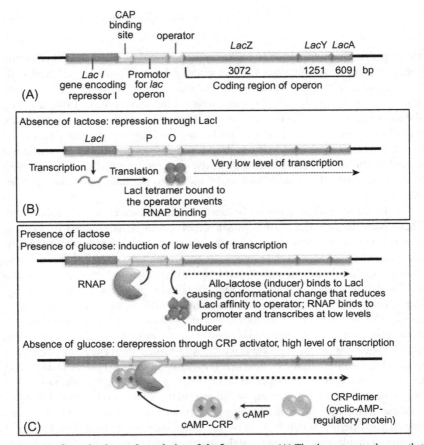

FIG. 11.7 Organization and regulation of the Lac operon. (A) The three structural genes that code for proteins required to metabolize lactose are: *LacZ* (codes for β-galactosidase, the enzyme that hydrolyzes the glycoside bond to split lactose into glucose and galactose), *LacY* (codes for the permease that transports the sugar from the medium through the cell membrane), and *LacA* (codes for the transacetylase, an enzyme which acetylates β-galactosides that cannot be hydrolyzed; this reaction eliminates toxic compounds from cell). The control region contains the promoter, the operator (which binds the Lac repressor), and the gene (with its promoter) that encodes the repressor molecule. The transcriptional activity of the lactose operon is regulated by the presence or absence of two sugars in the medium: lactose and glucose. (B) When glucose is present in the medium (independently of the presence or absence of lactose) the *Lac* operon is repressed, as glucose is the preferred carbon source. The presence of glucose leads to so-called catabolic repression that affects the activity of all operons that encode enzymes that catabolize alternative sugars. For the *Lac* operon to be active, two conditions must be met: lactose should be present, but glucose should not. (B) In the absence of lactose, the *Lac* operon is repressed: the LacI repressor is bound to a site that partially overlaps the *Lac* promoter, preventing the binding of RNAP. (C) The behavior of the operon in the presence of lactose. Upper panel, in the presence of glucose, the genes are only slightly expressed. A secondary lactose metabolite, 1,6-allo-lactose, binds to the repressor, changing its conformation and reducing its affinity for the operator. Allo-lactose acts as an inducer. For full activation to take place there should be no glucose in the medium. The mechanism of activation (bottom panel) in the absence of glucose involves an activator, catabolite activator protein (CAP) (also known as cyclic-AMP regulatory protein, CRP), which binds to a regulatory site near (upstream) of the promoter only if cAMP is bound to it. The absence of glucose is sensed by the cell through the intracellular concentration of cAMP: when glucose is high, cAMP levels are low; when glucose levels are low, there is high concentration of cAMP. cAMP then binds to CRP to change its conformation and facilitate its binding to DNA.

OVERVIEW

The aim of this chapter is to outline the thinking and research that finally clarified the path of information transfer from DNA to protein as postulated in the central dogma. In doing so, it has been necessary to introduce so many new concepts and structures that the overall picture tends to be lost. Basically, we are dealing with two dynamic processes: **transcription** to copy a DNA sequence into RNA, and **translation** to yield from the RNA a defined protein sequence. The greatly oversimplified pictures in Figs. 11.8 and 11.9 present the essence of these processes. In the last half century, it has been possible to examine these in almost molecular detail. This in no way diminishes the remarkable scientists of the 20th century, who laid such a firm groundwork.

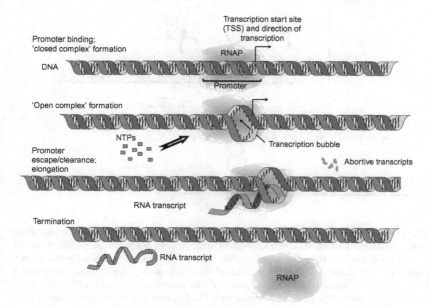

FIG. 11.8   The basics of transcription. The steps in transcription. The RNA polymerase (RNAP) binds to the promoter region on the DNA to form a "closed complex." RNAP melts the DNA to form the "open complex" in which the DNA strands separate to form a transcription bubble of 13–14 bp. The incoming ribonucleotide triphosphates pair with the exposed DNA bases on the template strand, and the first phosphodiester bonds are formed linking nucleotides within nascent RNA molecules. At this stage, RNA synthesis is abortive, and most newly formed short transcripts are released from the polymerase. Once a transcript elongates beyond ~15 bases, the polymerase clears the promoter and enters the processive elongation phase, in which the DNA-RNA-polymerase "ternary" complex is very stable and the enzyme transcribes long stretches of DNA without dissociating. The characteristic high processivity of RNA polymerase distinguishes it from other DNA-tracking enzymes, such as DNA polymerases and most helicases. Finally, transcription is terminated, resulting in dissociation of the ternary complex. *(Schematic courtesy of Andrei Revyakin, Janelia Farm Research Campus, Howard Hughes Medical Institute, Ashburn, VA, United States.)*

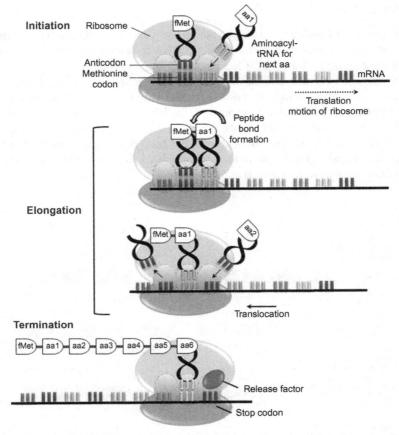

FIG. 11.9 The basics of translation. A schematic of the ribosome, with its three sites *(blue ovals)* that interact with the three tRNAs involved in the process. Initiation: special initiator tRNA carrying the amino acid formylmethionine (fMet) binds to the P site of the ribosome that accommodates the start codon. A new amino acid (aa)-tRNA then joins the complex, entering into the A site. The incoming amino acid must be cognate, i.e., the anticodon on the carrier tRNA must correspond to the codon in the message, to ensure the truthful transmission of the information encoded in the mRNA into the sequence of amino acids in the polypeptide chain. Elongation: peptide bond formation and translocation of the ribosome (with its bound tRNAs) with respect to the mRNA so that the next codon is now in the A site, ready to accept the next amino acid specified by the mRNA codon. Termination: when a stop codon enters the A site, translation is terminated with the help of special release factors.

EPILOGUE

The advances in molecular biology between 1953 and1963 were so great as to constitute a transformation. At the beginning of that decade the new science hardly existed; it was at best a collection of half-comprehended concepts. By the end, it possessed a comprehensive—if still incomplete—view of the molecular processes by which genetic information is stored, replicated, and transmitted for protein synthesis. Yet it must be remembered that almost all of the critical experiments had been conducted with bacteria or their viruses. Most of biology and medicine are concerned with the multicell organisms like ourselves: the eukaryotes. How much of what had been learned about the microscopic world would carry over?

FURTHER READING

Books and Reviews
Jacob, F., Monod, J., 1961. Genetic regulatory mechanisms in the synthesis of proteins. J. Mol. Biol. 3, 318–358. Covers not only the messenger hypothesis but also much of the *lac* studies.

Judson, H.F., 1979. The Eighth Day of Creation: Makers of the Revolution of Biology. Simon & Schuster, New York, NY. pp. 428–434. A detailed account of the fateful meeting at Brenner's. The whole book is a remarkably deep, but disorganized account of the rise of molecular biology.

Experimental Papers
Brenner, S., Jacob, F., Meselson, M., 1961. An unstable intermediate carrying information from genes to ribosomes for protein synthesis. Nature 190, 376–385. Experimental verification that new RNA uses old ribosomes.

Hall, D.B., Spiegelman, S., 1961. Sequence complementarity of T2 DNA and T2-specific RNA. Proc. Natl. Acad. Sci. USA 47, 137–146. Just what the title says.

Hoagland, M.B., Stephenson, M.L., Scott, J.F., Hecht, L.I., Zamecnik, P.C., 1958. A soluble ribonucleic acid intermediate in protein synthesis. J. Biol. Chem. 231, 241–257. Shows that RNA of a particular fraction of the cytoplasm becomes labelled with amino acids and subsequently transfers the attached amino acid to proteins on ribosomes.

Pardee, A., Jacob, F., Monod, J., 1959. The genetic control and cytoplasmic expression of "inducibility" in the synthesis of beta galactosidase by *E. coli*. J. Biol. Chem. 1, 165–178. The PaJaMo experiment.

Volkin, E., Astrachan, I., Countryman, J.L., 1958. Metabolism of RNA phosphorus in *Escherichia coli* infected with bacteriophage T7. Virology 6, 545–565. An early, uncited suggestion for a messenger RNA. Papers in two previous years provide background.

Zamecnik, P., Keller, E.B., 1954. Relation between phosphate energy donors and incorporation of labelled amino acids into proteins. J. Biol. Chem. 209, 337–353. An 'early' but important study.

Chapter 12

Eukaryotes Pose New Problems

PROLOGUE

Almost all of the advances in molecular biology before 1970 had utilized bacteria or bacteriophage as models for research. There were a number of good reasons for this. These were easy to work with, economical and reproduced rapidly. Their genomes were relatively small. Nevertheless, most scientists realized that studying the one-celled bacteria or their phage could not lead directly to the understanding of genetics and biochemistry of complex organisms like us. We are multicellular **eukaryotes**, with diverse cell and tissue types in each organism. In describing the multitude of research advances since 1970, it is necessary to first explain just what a eukaryote is, and how it differs from a **prokaryote**, like a bacterium.

WHAT IS A EUKARYOTE?

All living organisms on Earth consist of cells; the cell is the basic structural and functional unit of life. The cell was discovered by Robert Hooke in 1665, who studied a thin slice of cork under a simple microscope. He wittingly named the observed structures cells (*cella*, Latin for small room), for their resemblance to cells inhabited by Christian monks in a monastery. In 1839, Matthias Schleiden and Theodor Schwann developed the cell theory, which is still valid today. In its essence, the cell theory states that *every* organism is composed of one or more cells. We now believe that every cell contains the hereditary information necessary for performing and regulating its functions. Transmitting this information from one cell generation to the next is the only way to maintain life on Earth. In other words, all cells come from preexisting cells, and there is presently no spontaneous emergence of living cells from nonliving matter. Nevertheless, cells somehow emerged on Earth at least 3.5 billion years ago.

All cells contain cytoplasm, a liquid-gel medium which is separated from the surrounding environment by membranes. In bacteria, the DNA is in the form of a circular molecule suspended in the cytoplasm. The predecessors of present day bacteria gave rise, during the course of evolution, to a much more complex cellular structure, that of **eukaryotes**. Eukaryotes, whether unicellular or multicellular, are cells characterized by the existence of compartments

The Evolution of Molecular Biology. https://doi.org/10.1016/B978-0-12-812917-3.00012-7

called **organelles**, encapsulated in membranes and performing different specialized functions. Major examples are the **nucleus**, which contains most of the DNA, **mitochondria**, the major sites of energy production for the cell, and **chloroplasts**, the sites of photosynthesis in plants. All are shown in Fig. 12.1. Eukaryotes derive their name from the presence of the nucleus (*karyon*, Greek for kernel), which is separated from the cytoplasm by a double-layer membrane structure and which contains the majority of the genes in the form of several linear DNA molecules. Fossil records indicate that eukaryotic cells developed approximately 1.6–2.1 billion years ago. Many evolutionary biologists consider the emergence of eukaryotes as "perhaps the most important and dramatic event in the history of life" (Mayr, 2001). It is now generally believed that eukaryotic cells gained organelles like mitochondria and chloroplasts via symbiotic uptake of free-living bacteria or blue-green algae by a primitive nucleated prokaryote. This view is supported by the observation that both mitochondria and chloroplasts have their own DNA which codes for some (but not all) of their proteins.

THE ORIGINS OF EUKARYOTES

But how did eukaryotic cells come into existence? As always in science, questions that do not allow direct experimental inquiry are difficult to answer and can only give rise to hypotheses. Two classes of hypotheses concerning the origin of eukaryotes have been proposed. In the so-called autogenous models, primitive eukaryotic cells containing a defined nucleus existed first and later acquired various organelles. This is postulated to have occurred through developing invaginations in the plasma membrane; these evolved to satisfy the need for higher surface area to facilitate more efficient exchange of substances with the environment. These invaginations differentiated in function, gradually giving rise to many of the separate cellular compartments that exist today. Later, the first mitochondria were acquired through endosymbiosis between these already rather complex cells and invading aerobic proteobacteria. The importance of acquiring mitochondria as the energy plant of a cell is reflected in the view that all the eukaryotic lineages that did not acquire mitochondria went extinct. Finally, chloroplasts came from another endosymbiotic event involving cyanobacteria, organisms that obtain their energy through photosynthesis. The autogenous hypothesis found solid support in recent (2016) proteomic studies, where proteins common to all complex organisms have been tracked down to reconstruct evolutionary relationships.

The second types of models, the chimeric models, state that, first, two prokaryotic cells merged—by either a physical fusion or endosymbiosis—to form the precursor eukaryotic cells. These models have several versions, but the most widely accepted one is the serial endosymbiotic theory, introduced by Lynn Margulis in 1967. It specifies the bacteria involved: a motile anaerobic bacterium and another type of bacterium cell living in highly acidic and hot environments (*crenrachaeon*). The acquisition of mitochondria and chloroplasts is assumed to occur by further, similar, endosymbiotic events. The work of Margulis has been highly praised. Historian Jan Sapp stated that "Lynn Margulis's name is

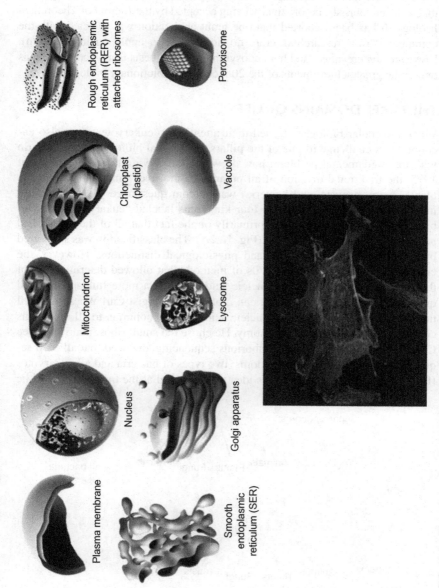

FIG. 12.1   The eukaryotic cell and its organelles. Schematic, three-dimensional representations of the constituents of eukaryotic cells. Chloroplasts are the photosynthetic organelles in plant cell. *Adapted from http://www.biology.arizona.edu/cell_bio/tutorials/pev/page3. html. (B) From Wikipedia.)*

as synonymous with symbiosis as Charles Darwin's is with evolution" (Sagan, 2012). Margulis story is one of courage and persistence in the face of skepticism and even derision. Her seminar work, published in 1967, was rejected by more than a dozen journals, before finally being accepted by the Journal of Theoretical Biology. It has been reported that one grant application was rejected with the comment: "Your research is crap, do not bother to apply again" (Ibidem). However, she prevailed, and her endosymbiotic hypothesis is now considered as one of the great achievements of the 20th century evolutionary biology.

THE THREE DOMAINS OF LIFE

When molecular biologists turned their attention to eukaryotes, they also encountered a challenge to one of the pillars of classical biology. This had to do with the fundamental problem: how do we classify living things? Until about 1975, the traditional division of all organisms into five "kingdoms" (animals, plants, fungi, protists, and bacteria) was seldom questioned. A super grouping was proposed, with the first four kingdoms labeled "eukaryotes," and all bacteria as "prokaryotes" based primarily on the fact that all of the former had nucleated cells, the latter did not (Fig. 12.2A). The classification was supported by anatomical, developmental, and physiological distinctions. However, the emergence in the 1960s and 1970s of methods that allowed determination of the sequences of proteins and nucleic acids allowed a more fundamental and quantitative approach. The American molecular biologist Carl Woese realized that comparing sequences of a nucleic acid that was common to all organisms could provide a more solid taxonomy. He chose the small ribosomal RNA (see Chapter 11), and after years of laborious sequencing, deduced that all such sequences fell into only three kingdoms; two types of bacteria and all eukaryotes (Fig. 12.2B; Table 12.1). Woese's idea was heretical at the time (1977) and he

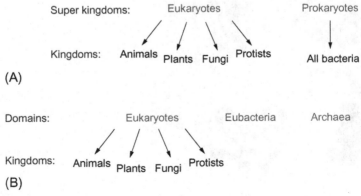

FIG. 12.2   Classification schemes for all existing organisms. (A) Classical classification based on morphological, developmental, and physiological features. (B) Newer classification based on sequencing data on nucleic acids and proteins.

TABLE 12.1 The Three Domains of Life

Domain	Cell Multiplicity	Major Organelles	Examples	Characteristics of Proteins, Information Transfer
Eubacteria (bacteria)	Unicellular	None	E. coli, most common bacteria	Different from eukaryotic; some similarities to archaea
Archaea	Unicellular	None	Methanogens; heat- and salt-tolerant microorganisms	Characteristics of both eubacteria and eukaryotae; some unique lipids
Eukaryotae (eukaryotes)	Unicellular (Protista) and multicellular (Metazoa)	Nucleus, mitochondria, chloroplasts (in plants)	Amoeba, human, apple tree	Generally different from eubacteria; some similarities to Archaea

From Molecular Biology: Structure and Dynamics of Genomes and Proteomes by Jordanka Zlatanova and Kensal E. van Holde. Reproduced by permission of Taylor and Francis Group, LLC, a division of Informa plc. Copyright 2015

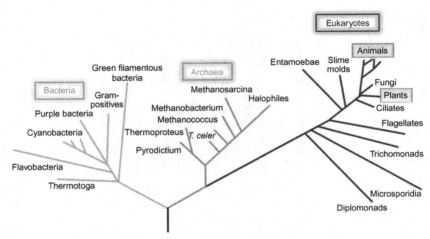

FIG. 12.3   The Tree of Life as deduced from molecular data. A common unknown ancestor at the base of the tree gave rise to three different branches (domains of life): bacteria, archaea, and eukaryotes. The lack of a known ancestor is designated by evolutionary biologists as a "rootless tree." The lengths of the branches reflect how much the DNA of each lineage has diverged from their common ancestor. The tree demonstrates that most of life's genetic diversity is in the microbial domain; the entire animal kingdom (in the old classification) is represented by just a few twigs at one end of the tree. Notably, multicellular organisms such as fungi, plants, and animals have all evolved from unicellular organisms further down the tree. *(Adapted from Woese, C.R., 2000. Interpreting the universal phylogenetic tree. Proc. Natl. Acad. Sci. USA 97, 8392–8396.)*

had to fight the rest of his life for it to gain acceptance. Such molecular data lead to a new tree of life (Fig. 12.3) which is now widely accepted. We humans occupy a very unimpressive place—at the tip of a small twig on a long branch.

INTERRUPTED MESSAGES AND SPLICING

By 1967, the fundamentals of molecular biology seemed to be firmly established, at least insofar as bacteria and viruses were concerned. The roles of messenger RNA, transfer RNA, and ribosomes in transmitting information from DNA to protein were clear. The code had been completely worked out and seemed universal for viruses and bacteria. The essential enzymology for the transcription of mRNA from DNA (RNA polymerase) was recognized.

When researchers then turned their attention to the molecular biology of eukaryotes, many surprises were in store. First, in eukaryotes, there turned out to be three RNA polymerase complexes, each with a different specialization in terms of specific types of genes they transcribe. This was in sharp contrast to the situation in bacteria, where only one type of RNA polymerase exists. Second, there were exceptions to the bacterial code in some lower eukaryotes and in mitochondria. But the great shock was the discovery, first by Phillip Sharp and Richard Roberts in 1977, that the mRNA found in the cytoplasm of eukaryotic cells was, in most cases, not an exact, complete copy of the corresponding gene in the nuclear DNA. Specifically, it was found that within the genes there were stretches of DNA sequence that had no counterparts in the cytoplasmic mRNAs. These intervening sequences in the genes, later called **introns**, did not correspond to any part of the protein sequence that was subsequently translated. They lay between sequences that did code for protein, now called **exons**. Roberts and Sharp were awarded the 1993 Nobel Prize in Physiology or Medicine "for their discoveries of split genes."

The introns first discovered by Roberts and Sharp lay in the genes of adenovirus, a virus that infects human respiratory cells, and so it was suspected at first that introns might be some idiosyncrasy of such viruses. This expectation was soon completely dashed when the French scientist Pierre Chambon demonstrated that the chicken ovalbumin gene contained no less than seven intervening sequences, along with eight exons (Fig. 12.4)! Furthermore, the total length of introns was far greater than the total length of the exons that comprised the cytoplasmic message. Soon, other eukaryotic genes were found to contain introns. We now know that eukaryotic genes that lack introns are in the minority.

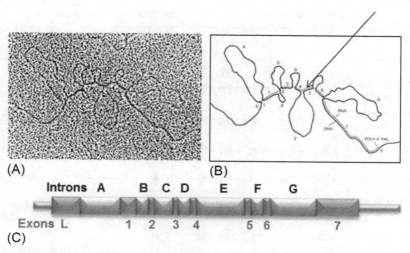

FIG. 12.4   The ovalbumin gene and its mRNA illustrate the concept of split genes. The protein is 386 amino acids long and could have been encoded by a gene of 1158 bp, but the length of the gene is actually 7700 bp. This anomaly was explained when ovalbumin mRNA was purified from the cytoplasm, allowed to hybridize with the ovalbumin gene, and the resulting hybrid was examined under the EM (A). (B) In this schematic interpretation of the EM image, the *black line* shows the portion of the gene that does not have complementary representation in the mRNA (introns). The mRNA-DNA hybrid is presented as a *gray line*. (C) A linear presentation of the ovalbumin gene structure, with the 7 introns (A–G) in *lighter gray* and the 8 exons in *darker gray*. The gene is transcribed as a single long pre-mRNA, which is then processed to give rise to the mature mRNA that serves as the template in protein synthesis. (*From Chambon, P., 1981. Scientific American, 245, 60–71, Fig. 1, with permission.*)

Clearly, something special must take place in the eukaryotic nucleus to prepare mRNA molecules for translation on the ribosome—the introns still present in the primary transcript of the gene must be removed. In subsequent years, an elegant process of **splicing** was discovered. The gene, it turns out, is first completely transcribed, introns and all, to make a **pre-mRNA**. This is precisely cut and resealed to make a transcript containing only exon regions, in the proper order. After some further adjustments to the 5'- and 3'-ends, the mature message is delivered to the cytoplasm for translation. The signals as to where to cut and splice are all contained in the pre-RNA sequence.

An interesting twist to the splicing story was the discovery that one and the same pre-mRNA molecule can be spliced in a number of alternative pathways, giving rise to multiple mature mRNAs, and hence a number of alternative protein forms (e.g., see Fig. 12.5). In humans, ~95% of pre-mRNA molecules that are transcribed from split genes undergo alternative splicing. Hence, eukaryotes have evolved a complex and highly regulated splicing machineries and mechanisms to produce the protein forms needed by particular cell types. Interesting and highly significant analyses showed that >60% of human disease-causing

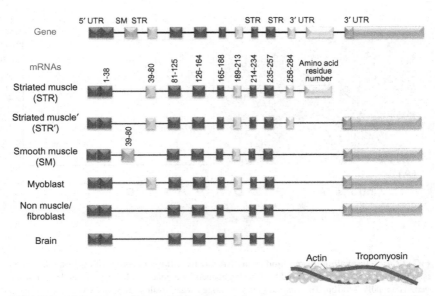

FIG. 12.5 Alternative splicing of the tropomyosin gene. Tropomyosin is an actin-binding protein that regulates the binding of myosin to actin in muscle fibers; this interaction is important in muscle contraction and other actomyosin functions. Distinct tropomyosin variants are expressed in different tissues, with some exons selectively present or absent in some tissue-specific isoforms. In addition to the common form of alternative splicing—exon skipping—the production of some isoforms involves the utilization of alternative polyadenylation sites (top and bottom variants). (*Adapted from Breitbart, R.E., et al., 1987. Annu. Rev. Biochem. 56, 467–495, Fig. 3, with permission.*)

mutations affect the functioning of the splicing machineries, rather than the coding sequences themselves.

EVERY CELL TYPE HAS SPECIAL NEEDS AND FUNCTIONS

Unicellular eukaryotic cells of a given species, such as yeast, are practically all the same. However, this is not the case with multicellular eukaryotic organisms. An average human body is estimated to have around 37 trillion cells, with around 200 different cell types. The different types have different needs and perform different, highly specialized functions. Thus, a liver cell is amazingly different from a brain cell, or a muscle cell. In addition, cells in a given organ can be highly heterogeneous in themselves. There are at least several cell types in blood: erythrocytes that transport oxygen, a variety of white blood cells (macrophages, T cells, neutrophils, etc.) that are involved in the immune response of the body, and platelets that are involved in blood clotting. The liver contains hepatocytes that break down and store lipids, carbohydrates, and amino acids. Hepatocytes also produce bile for fat digestion and remove toxic compounds from the body. Even these functions are not performed by all hepatocytes; the specific tasks depend on the exact location of the cells within the liver. In addition to hepatocytes, the liver contains Kupffer cells, stellate cells, and sinusoidal endothelial cells, each with a different function.

How all these cell types can be generated during development from a single fertilized egg (zygote) is a major and fascinating topic of research. We will discuss the methods used to approach differentiation and development in Chapter 13.

MULTIPLE LEVELS OF CONTROL

Every cell in a multicellular organism has the same DNA. Yet only a fraction of that DNA is expressed in each cell type, and what is expressed will differ from one cell type to another. For example, the bone marrow cells synthesize hemoglobin for the blood, but they do not make digestive enzymes like trypsin. That is the responsibility of pancreatic cells. Yet both types have the same DNA.

It is clear that the complexities of structure and function that characterize eukaryotic cells require complex, multilevel control systems. These are especially evident from recent genome-wide studies of protein gene localization along DNA molecules, global analysis of the transcriptome (the set of all genes transcribed in a certain cell type under certain conditions), analysis of the multiple types of small and long noncoding RNA molecules, and similar lines of research. While it is beyond the scope of this book to describe these mechanisms in detail, we feel that appreciation of these multiple levels of control is important. Furthermore, many of the major advances in molecular biology from 1970 until the present day have dealt with the organization of eukaryotic genomes and control of their expression. We begin with a general level of control that is shared by all eukaryotes.

CHROMATIN AND NUCLEOSOMES

Recall from Chapter 6 that the material Miescher extracted from human cells in 1868 contained both nucleic acid and proteins. That was no accident of primitive technique. As demonstrated in 1884 by Albrecht Kossel, the nucleic acids in eukaryotic nuclei are complexed with specific proteins, which he termed **histones**. The complex was soon given the name **chromatin**, derived from the fact that this material could be stained in the nucleus by certain dyes. It is important to note that bacteria do not have this complex. Their DNA resides in the cytoplasm, and although it interacts with a number of proteins, these are not histones. Chromatin is unique to eukaryotic cells.

Because there is only a small set of histones common to all eukaryotes, and because they constitute by far the most abundant proteins in the nucleus, it was long suspected that histones might play a role in the compaction and regulation of DNA. Both functions are essential; only a portion of the eukaryotic genome is read to dictate protein or RNA sequences, and the compaction necessary to squeeze a meter of human DNA into a micrometer-size nucleus is impressive. But how do histones do it?

The answers did not come until the early 1970s—in fact, there is still much to understand. A number of laboratories, using techniques ranging from nuclease digestion to electron microscopy to ultracentrifugation to chemical protein-protein cross-linking, and finally X-ray diffraction converged to yield the model shown in Fig. 12.6. A complex of eight histone molecules, two each of histone classes H2A, H2B, H3, and H4, forms a protein core about which about 147 bp of DNA is helically wrapped. These **nucleosomes** are spaced more or less regularly along the eukaryotic DNA. This accomplishes considerable compaction, which is further enhanced by condensation of the chromatin fiber, as shown in Fig. 12.7. The local degree of condensation can also provide regulation as to which genes are expressed or repressed. For example, high-resolution nucleosome mapping in the **ENCODE** project (see below) showed that transcription start sites generally correspond to locally nucleosome-free regions. However, the whole picture is enormously more complicated, involving various kinds of chemical modifications to the histones themselves, which in turn modulate binding of other cellular proteins to the chromatin fiber (Fig. 12.8). There are modifications to some of the bases in DNA nucleotides too, the most prominent being methylation (addition of a CH_3 group to cytosine residues). Because such modifications change the reading of the genome despite the dictates of genetics, they are referred to as **epigenetic**.

The recognition of the important role of epigenetic modifications in regulation of gene expression has led to numerous studies of these modifications in disease state. A prominent example is human cancer. Hypermethylation of cytosine residues in promoter regions of tumor-suppressor genes (genes whose function is to protect normal cells from turning cancerous) silences these genes. This DNA modification is associated with particular alterations of the "normal"

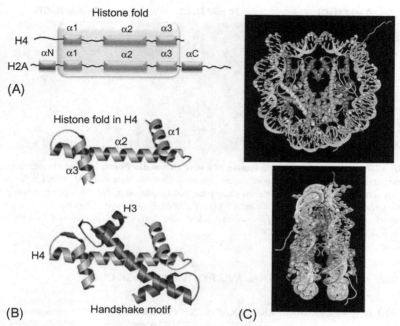

FIG. 12.6   Crystal structure of the core nucleosome at 2.8Å resolution. (A) All four core histones share a common structural motif, called the **histone fold**. This fold consists of three α-helices arranged as shown in the crystal structure of histone H4, when H4 is part of the core particle. The individual histones differ in the presence or absence of additional α-helices at one or both ends of the polypeptide chain. Thus, histone H2A possesses additional α-helices at both ends of the molecule (αN and αC). (B) The individual histone molecules interact with each other along their long α2 helices in what is known as the **handshake motif**. The structures of the H2A-H2B dimers and H3-H4 dimers are almost superimposable. (C) Two views perpendicular to the twofold axis of symmetry of the core particle (known as the dyad) are shown. Color scheme: H3, *blue*; H4, *green*; H2A, *yellow*; H2B, *red*. The main technical difficulty in obtaining this high-resolution structure was to get good crystals. To reduce the inherent heterogeneity in particles isolated from nuclei, it was crucial to use recombinant histones that lack posttranslational modifications and are devoid of the N-terminal tails which seem highly disorganized. The tails in the structure show the presumptive locations of some of the tails. In addition, the DNA sequence used for reconstitution was constructed of two identical halves, connected head-to-tail. *((B) Reproduced with permission from Harp, J.M., et al., 2000. Acta Crystallogr. D56, 1513–1534, Figs. 6 and 7, with permission. (C) From Luger, K., Mader, A.W., Richmond, R.K., Sargent, D.F., Richmond, T.J., 1997. Crystal structure of the nucleosome core particle at 2.8 A resolution. Nature 389, 251–260, Fig. 1A, with permission.)*

histone modification patterns: histones H3 and H4 are deacetylated and certain lysine residues in histone H3 undergo specific changes in their methylation patterns, either losing or gaining methyl groups. Interestingly and importantly, the presence of the hypoacetylated and hypermethylated histones H3 and H4 silences certain tumor-suppressor genes, even in the absence of DNA hypermethylation of their promoter regions. In human tumors, generally, modifications

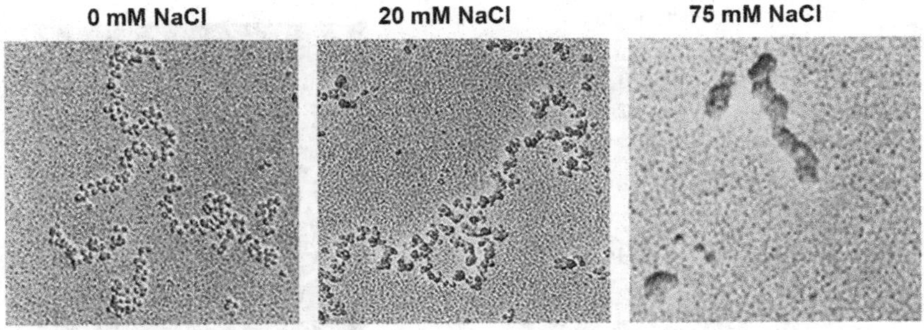

0 mM NaCl   **20 mM NaCl**   **75 mM NaCl**

FIG. 12.7   Electron microscope images of native chromatin fibers. Isolated chromatin fibers were fixed with glutaraldehyde under various salt concentrations, then dialyzed against buffers that do not contain salt and imaged. One can clearly see the compaction of the fiber upon increasing the ionic strength. *(From Thoma, F., Koller, T., Klug, A., 1979. Involvement of histone H1 in the organization of the nucleosome and of the salt-dependent superstructures of chromatin. J. Cell Biol. 83, 403–427, Fig. 3, with permission.)*

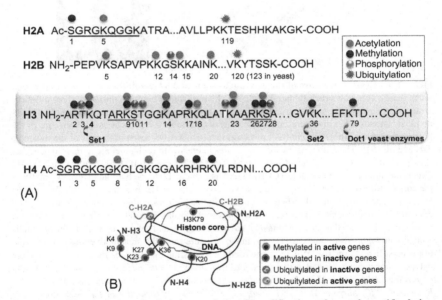

FIG. 12.8   Histone postsynthetic (posttranslational) modifications along polypeptide chains of the four core histones. (A) Underlined sequences are motifs that are repeated either in two different molecules, or within the same molecule. The significance of having such repeats is unclear. The best studied modifications are those of histone H3. The enzymes that place the same modification on different residues are environment specific: thus, in yeast, different enzymes, Set1, Set2, and Dot1, modify lysines 4, 36, and 79, respectively. (B) Most posttranslational modifications occur in the histone tails that protrude from the core particle structure; however, modified residues also exist within the histone fold, which is entirely inside the particle. For clarity, only half of the core histones in the particle are shown. The schematic also illustrates which modifications are present in chromatin regions that are actively transcribed or which are transcriptionally inactive. For example, histone acetylation at lysine residues is a characteristic feature of active genes.

of histone H4 entail a loss of monoacetylated and trimethylated forms. These changes appear early and accumulate during the development of the tumor.

Elucidation of chromatin structure and its relation to genome expression was an essential step in opening molecular biology into the vastly important realm of eukaryotic organisms—especially us, human beings.

TOO MUCH DNA? JUNK DNA?

For years, the function of the genomic DNA was viewed only in terms of coding for proteins and some functional RNA molecules, such as ribosomal and transfer RNA. Sequences that regulate the expression of such genes have also been recognized, both upstream (to the 5′-end) and downstream (to the 3′-end) to the gene. Yet there was still a relatively large amount of DNA that was seemingly not involved in any of these functions. Sequencing of whole genomes from a variety of organisms, both bacteria and eukaryotes, gave precise numbers for the sizes of their genomes (Fig. 12.9). Remarkably, the amount of DNA involved in coding for proteins is relatively uniform over a wide evolutionary span; on the other hand, the total DNA per genome varies enormously. Something is strange! For a number of years, the concept of "**junk DNA**"—DNA which had no function—was popular. But new data discredit this concept.

The major contribution to our present understanding of eukaryotic genomes and their functional organization has come from a total analysis of the human

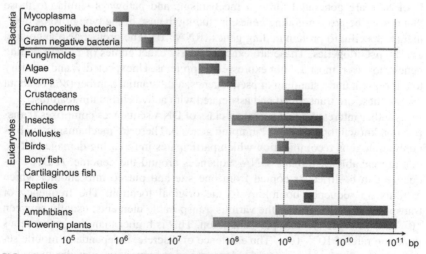

FIG. 12.9   Variation in genome sizes among different taxa. Note the tremendous variation of genome sizes among eukaryotes in general, and the relatively small range of variation among mammals *(dark gray box)*. Based on completely sequenced genome data (as of mid-2009), gene number correlates linearly with genome size in viruses, bacteria, archaea, and eukaryotic organelles, whereas there is no apparent correlation in eukaryotes. Moreover, the number of genes does not correlate with the complexity of the organism.

genome, performed by the **Encyclopedia of DNA Elements (ENCODE) project** consortium. This consortium was created in 2003 under the auspices of the National Human Genome Research Institute in the United States and consists of a large international group of scientists with broad expertise in genome-wide experimental methods and data analyses using computational methods. The consortium operationally defines a **functional element** as a "discrete genome segment that encodes a defined product (e.g., protein or noncoding RNA) or displays a reproducible biochemical signature (e.g., protein binding or a specific chromatin structure)" (The ENCODE Project Consortium, 2012). The elements mapped across the human genome include all regions transcribed into RNA, protein-coding regions, transcription-factor binding sites, features of chromatin structure, and sites where DNA is chemically modified, by, for instance, adding a methyl group to cytosine residues (see above).

ENCODE identified 20,687 protein-coding genes (with an indication that some additional protein-coding genes remain to be found). These genes constitute less than 3% of the entire genome! This proportion is much smaller than had been expected and made it impossible to continue to think that the major role of DNA is to code for proteins, especially since ENCODE data indicate that around 80% of the entire human genome is transcribed into *something*! A very large fraction of the human genome is, instead of protein-coding, being transcribed into a number of classes of nonmessenger RNA molecules, most of presently unknown function. The number of small (<200-nucleotide) RNA molecules has been estimated at 8801, and that of long noncoding RNAs (lncRNAs) at 9640. LncRNAs are generated through mechanisms and pathways similar to those that transcribe protein-coding genes, including the use of the type of RNA polymerase specific to protein-coding genes (RNAP II). The project also annotated 11,224 **pseudogenes**. These are exact or nearly exact copies of protein-coding genes; however, most are not expressed as proteins. Unexpectedly, and contrary to our present understanding of pseudogenes, a substantial number (863 or about 7%) of these are transcribed and associated with active chromatin features.

Another interesting and prevalent class of DNA sequences comprises **transposons** (known for a while as "jumping genes"). There are mechanisms, such as nonhomologous recombination which participates in repairing damaged DNA, that are capable of moving DNA sequences around the genome. A DNA sequence can be lifted or copied from one site and placed into another which exhibits no sequence homology to the original location. The frequency of transposition varies among the various transposable elements, usually between 10^{-3} and 10^{-4} per element per generation. This is higher than the spontaneous mutation rate of 10^{-5}–10^{-7}. The existence of discrete, independent, mobile sequences, known as **transposable elements** or **transposons**, was discovered in maize (*Zea mays*) in 1948 by Barbara McClintock. At the time, her visionary work was accepted with much skepticism. Eventually, recognition came in the form of the 1983 Nobel Prize in Physiology or Medicine for "her discovery of mobile genetic elements."

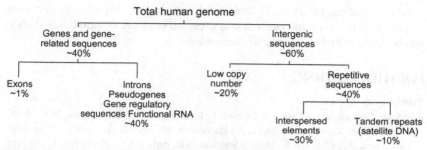

FIG. 12.10   An approximate distribution of DNA functions in the human genome. Note that
the DNA that actually codes for proteins is a very small fraction of the genome. The function of
the low-copy-number intergenic sequences is not well understood. The interspersed elements are
mostly transposons. The satellite DNA can be further subdivided into minisatellites (repeats of
10–15 bp, repeated up to ~1000-fold) and microsatellites (2–6 bp, up to 100 copies).

Transposons are abundant, scattered throughout the genomes of many plants
and animals and can constitute a considerable portion of the DNA. Thus, in
mammals, transposons occupy ~40%–45% of the genome; the percentage is
even higher in some plants (~60% in maize). But the largest fraction known
is in the frog *Rana esculenta* (77%). The reasons for these differences are not
understood. Transposons are also present in bacteria and unicellular eukaryotes
but comprise a much smaller portion of the genome (transposons in *Escherichia
coli* constitute ~0.3% of the genome, in *Saccharomyces cerevisiae*, 3%–5%).

Roughly 10% of the whole human genome consists of tandem repeats, also
referred to as **satellite DNAs**. Satellite DNA is so named because it can be
identified as "satellite" bands, separated from the bulk DNA, in density gradient
equilibrium sedimentation (This is the technique that was used in the Meselson-
Stahl experiment) (see Chapter 8 and Fig. 8.2). If whole DNA is cleaved with
a restriction nuclease that does not cut in the repetitive sequence, the stretches
of repeats produced will often have an overall base composition very different
from the bulk [consider an $(AT)_n$ repeat, for example]. DNA density depends on
base composition, so such DNA will band separately from the bulk. Individual
humans have distinctive patterns of such repeats; analysis of microsatellites is
being widely used in forensics (see Chapter 17).

The approximate distribution of DNA functions in the human genome is
presented in Fig. 12.10.

EPILOGUE

Science is full of surprises. Many molecular biologists, looking to extend the
pioneering work on phage and bacteria to the eukaryotes, expected just more of
the same, with minor additions. They were wholly unprepared for the prospect
that molecular biological methods might rewrite taxonomy. The existence of introns and DNA splicing were completely unexpected shocks. "Jumping genes"

were not believed for years. If there is a lesson in this chapter, it is that there is more under heaven and earth than is dreamed of in our complacency. This is much of which makes science so fascinating.

FURTHER READING

Books and Reviews

Adl, S.M., Simpson, A.G., Lane, C.E., Lukes, J., Bass, D., Bowser, S.S., Brown, M.W., Burki, F., Dunthorn, M., Hampl, V., Heiss, A., Hoppenrath, M., Lara, E., Le Gall, L., Lynn, D.H., McManus, H., Mitchell, E.A., Mozley-Stanridge, S.E., Parfrey, L.W., Pawlowski, J., Rueckert, S., Shadwick, L., Schoch, C.L., Smirnov, A., Spiegel, F.W., 2012. The revised classification of eukaryotes. J. Eukaryot. Microbiol. 59, 429–493. A global tree of eukaryotes from a consensus of phylogenetic evidence, rare genomic signatures, and morphological characteristics.

Keeling, P.J., Koonin, E.V. (Eds.), 2014. Origin and Evolution of Eukaryotes. first ed. Cold Spring Harbor Laboratory Press, Long Island, NY. A summary of conflicting theories.

Mayr, E., 2001. What Evolution Is. Basic Books, New York, NY. Presents evidence for evolution of life on Earth. Extensive glossary and bibliography.

Merchant, R.G., Favor, L.J., 2015. How Eukaryotic and Prokaryotic Cells Differ. Britannica Educational Publishing and The Rosen Publishing Group, New York, NY. A popular expose of the differences between eukaryotic and prokaryotic cells.

Sagan, D., 2012. In: Lynn Margulis: The Life and Legacy of a Scientific Rebel. Chelsea Green Publishing, White River Junction, VT. A biography of an unusual scientist by a major coauthor.

Sagan, L., 1967. On the origin of mitosing cells. J. Theor. Biol. 14, 255–274. A theoretical paper addressing the discontinuity between the modes of division of bacterial and eukaryotic cells based on data on the biochemistry and cytology of subcellular organelles.

Schubeler, D., 2009. Epigenomics: methylation matters. Nature 462, 296–297. Comment on the significance of the first complete DNA-methylation map of the human genome at single-base-pair resolution. Distinct differences among cell types were revealed.

van Holde, K.E., 1988. Chromatin. Springer Verlag, New York, NY. A comprehensive view of chromatin research until about 1986.

van Holde, K., Zlatanova, J., 2007. Chromatin fiber structure: where is the problem now? Semin. Cell Dev. Biol. 18, 651–658. Conflicting views on chromatin fiber folding.

Willbanks, A., Leary, M., Greenshields, M., Tyminski, C., Heerboth, S., Lapinska, K., Haskins, K., Sarkar, S., 2016. The evolution of epigenetics: from prokaryotes to humans and its biological consequences. Genet. Epigenet. 8, 25–36.

Classic Research Papers

Luger, K., Mader, A.W., Richmond, R.K., Sargent, D.F., Richmond, T.J., 1997. Crystal structure of the nucleosome core particle at 2.8 A resolution. Nature 389, 251–260. The first high-resolution X-ray diffraction study of the nucleosome.

The ENCODE Project Consortium, 2012. An Integrated encyclopedia of DNA elements in the human genome. Nature 489, 57–74. A major analysis of genome-wide data on the human genome.

Thoma, F., Koller, T., Klug, A., 1979. Involvement of histone H1 in the organization of the nucleosome and of the salt-dependent superstructures of chromatin. J. Cell Biol. 83, 403–427. A systematic study, using electron microscopy imaging, of the effects of ionic strength on the morphology of chromatin fibers and of fibers depleted of the linker histone H1.

Woese, C.R., 2000. Interpreting the universal phylogenetic tree. Proc. Natl. Acad. Sci. USA 97, 8392–8396. The fundamental statement of his work, underlying the importance of trees based on rRNA sequences. Such trees transcend the era of modern cells, extending back to times when cells were still very primitive.

Chapter 13

Development and Differentiation

PROLOGUE

The question of how human embryos come into existence and further develop to form a fully functional organism has fascinated both scientists and the general public for millennia. Only in the past few centuries have we begun to understand how embryonic and postembryonic development occurs; how a single fertilized egg can give rise to the complexities of a multicellular organism, in which hundreds of different cell types with specialized functions exist and communicate with each other. Here, we will provide a bit of historical background on these issues and will then focus on critical approaches and experiments, performed during the later decades of the 20th century, particularly those inspired by the advent of molecular biological approaches and insights. This is an example of where a new field can spill over into an old one and revitalize it.

TWO IDEAS ABOUT DEVELOPMENT DOMINATED THINKING IN ANCIENT TIMES

The Greek philosopher Aristotle (384–322 BC) has been considered by some to be "the first genuine scientist in history" (*Encyclopaedia Britannica*, The Britannica guide to the 100 most influential scientists). Aristotle described the two competing ideas of embryonic development. In the model of **preformation**, an embryo is a miniature individual (homunculus) that preexists in either the mother's egg or the father's sperm. This preformed homunculus simply begins to grow in utero, becoming larger and larger during development. Some preformationists also believed that all the embryos that would ever develop had been formed by God at the Creation. The alternative theory, the one actually favored by Aristotle, was that of **epigenesis**. Epigenesis assumes that that early in life, the embryo begins as an undifferentiated mass and that new, morphologically distinguishable features appear later in development. This is quite close to current knowledge, except Aristotle believed that the egg contributed by the mother provided the unorganized matter, which was then provided a "form-building principle" by the semen of the male parent.

The Evolution of Molecular Biology. https://doi.org/10.1016/B978-0-12-812917-3.00013-9

THE INTRODUCTION OF SCIENTIFIC APPROACHES TO THE FIELD OF DEVELOPMENT

The first to put embryology on solid experimental grounds was the remarkable British scientist William Harvey, most famous for having discovered the circulation of the blood. Harvey dissected the uterus of female deer at time intervals following mating and was unable to find any signs of preformed body parts until 6 or 7 weeks following mating. Similar results were found with chicken eggs. His book *On the Generation of Animals*, published in 1651, convincingly supported the notion of epigenesis. Despite this, many scientists were still supporting the preformation theories as recently as the 18th century. The development of the microscope contributed much to the downfall of preformationism (although van Leuwenhoek himself still maintained it, at least in part). A critical critique, based on careful studies of actual embryos, was presented in the Ph.D. thesis of Caspar Wolff at the Academy of St. Petersburg in 1759. Despite all of the solid early evidence against it, preformationism continued to find adherents until well into the 19th century.

AN OPPORTUNITY MISSED?

In 1935, a proposal was presented to the Rockefeller Foundation to establish an "Institute of Mathematical and Physico-Chemical Morphology" at Cambridge University, United Kingdom. The proposed staff included some forward-looking scientists, like the crystallographer John D. Bernal, and the embryologists Joseph Needham and Conrad Waddington. If it had been founded with just such people, the history of molecular biology might have been different. Instead, most interest in the proposal was devoted to finding mathematical models for development, an approach that has never proved fruitful. Warren Weaver, the director of the Rockefeller program, wisely rejected the plan. Embryology had to wait another three decades for rebirth.

WHAT DO WE KNOW ABOUT DEVELOPMENT AND DIFFERENTIATION AT PRESENT?

It is now clear that development begins with fertilization, the process of formation of a **zygote** through unification of haploid egg and sperms cells. The resulting diploid zygote gives rise to the numerous (~220) differentiated cell types in an adult human, each of which is specialized in performing unique functions. These functions are performed by a variety of protein molecules. Some proteins have "housekeeping" functions; they are needed for the essential activities in all cells; some are highly specialized and are uniquely present in only one or a few highly specialized cell types.

It is also accepted that development is a unidirectional process, that is, with each successive cell division the cell loses its ability to give rise to more types of cells. Developmental biologists now discriminate between different levels of "potency." **Totipotency** describes the ability of a given cell to give rise to all cells of an organism, including embryonic and extraembryonic tissues: fertilized eggs (zygotes) are totipotent. **Pluripotency** refers to the ability to produce all cells of the embryo proper; cells of the inner cell mass are pluripotent, so are the **embryonic stem cells** (ESC) derived from these cells and grown in culture (see below). We also refer to **multipotency,** describing the developmental potential of cells capable of giving rise to different cell types of a given cell lineage: most adult stem cells are multipotent. These include skin stem cells, hematopoietic (blood forming), and neural stem cells. Finally, terminally differentiated cells and committed progenitor cells, such as erythroblast (immature erythrocytes), are described as **unipotent.** These views of unidirectionality of the developmental process are illustrated by the "epigenetic landscape model" introduced by Conrad Hal Waddington in 1957 (Fig. 13.1).

ESC SERVE AS A MODEL FOR PLURIPOTENCY

ESC are derived from the inner cell mass of a dividing embryo and cultivated in nutrient media in laboratory vessels (Fig. 13.2). They remain pluripotent, as the embryo cells they originate from. Thus, they are self-renewing and can give rise, upon appropriate treatment, to all lineages and cell types of an adult organism.

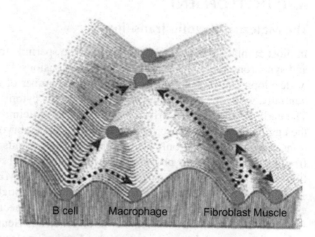

Developmental
potential

Totipotent
(zygote)

Pluripotent
(many cell types,
including embryonic
stem cells)

Multipotent
(adult stem cells)

Unipotent
(differentiated cell
types)

B cell   Macrophage   Fibroblast Muscle

FIG. 13.1   The developmental potential and epigenetic states of cells at different stages of development. A modification of C. H. Waddington's epigenetic landscape model, showing cell populations with different developmental potentials. Developmental restrictions can be illustrated as marbles rolling down a landscape into one of several valleys (cell fates). Marbles correspond to different differentiation states (from top to bottom: totipotent; pluripotent, multipotent, and unipotent). Examples of reprogramming processes are shown by *dashed arrows*. (*Adapted from Hochedlinger, K., Plath, K., 2009. Epigenetic reprogramming and induced pluripotency. Development 136, 509–523, with permission.*)

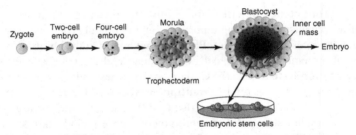

FIG. 13.2 Schematic depicting how embryonic stem cells (ESC) are derived. ESC are derived from the inner cell mass of a dividing embryo at the blastocyst stage and cultivated in the laboratory in nutrient media.

This potential provides researchers with a powerful tool to study differentiation and development in the test tube.

The ability of ESC to develop into an entire organism has created ethical issues. The potential to generate cells, tissues, and organs for clinical practice is no doubt useful. On the other hand, though, ESC can be seen as dangerous human cloning which is prohibited by law in most developed countries. By contrast, cloning of animals and plants has been accepted as a way to improve food production (see Chapter 17).

THE MOLECULAR BASIS OF DIFFERENTIATION AND DEVELOPMENT

The Maternal-Zygotic Transition

In most animals, fertilization of the egg by the sperm is followed by very rapid and synchronous cell cycles, called cleavage divisions. During this early period of development, the amount of DNA and the number of nuclei increase exponentially, while the total amount of the embryonic cytoplasm remains constant. The early development is driven by mRNA and protein molecules provided by the egg, with no new transcription occurring. The embryonic genome becomes transcriptionally active only later. The molecular events during the transition from maternal to zygotic phases of development are well studied, and the transition itself is known as the maternal-zygotic transition (MZT) (Fig. 13.3).

Maternal RNAs undergo controlled degradation through normal mechanisms employed throughout adult life, including binding of specific proteins and small RNAs to the 3′ untranslated regions of target RNA. Genome activation relies, at least in part, on the sequestration (binding) of molecules that repress early transcription by the huge amount of DNA replicated during the cleavage divisions. Multiple, nonmutually exclusive models have been suggested to explain the complex network of molecular events that take place during the MZT. The transcriptionally inactive genome following fertilization is activated to give rise to cells in the blastocyst, which are pluripotent. In the subsequent development of the embryo levels of potency are successively lost (Fig. 13.4).

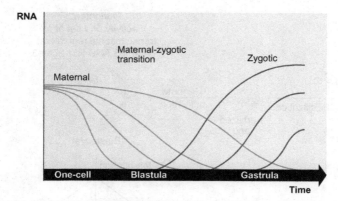

FIG. 13.3   The maternal-zygotic transition. The embryonic genome is initially transcriptionally inactive; zygotic genes start to be transcribed only during this transition phase. All proteins in the fertilized egg are translated from maternal RNAs, which are deposited by the mother into the egg and drive early development. These RNAs, shown by different curves, are degraded during different stages of embryogenesis, including blastula and gastrula stages. *(Adapted from Schier, A. F., 2007. Science 316, 406–407, with permission.)*

Genes Control Development: The Case for the Fruit Fly

The early search for genes that control early development and differentiation made use of the model organism the fruit fly, *Drosophila melanogaster*. The fruit fly has many characteristics that make it a highly desirable model for genetic and now molecular biology research (see Chapter 5).

The fly has only four pairs of chromosomes and the nucleotide sequence of the entire genome was published in 2000. The genome contains ~15,500 genes; ~60% of the genome does not code for proteins and is involved in the control of gene expression. Finally, techniques for genetic transformation have been available since 1987, expanding the usefulness of the fly as an experimental organism.

Of significant practical interest is the fact that ~60% of the *Drosophila* genes are conserved in humans, including many known human disease genes. This conservation has led to the widespread usage of *Drosophila* as a genetic model for several human diseases including neurodegenerative disorders such as Parkinson's, Huntington's, spinocerebellar ataxia, and Alzheimer's disease. The fly is also being used to study basic molecular mechanisms underlying aging and oxidative stress, immunity, diabetes, and cancer.

The embryonic development of the fly is quite well understood, again providing a powerful system to study development and differentiation. After fertilization, rapid DNA replication cycles take place in the early embryo, followed by numerous divisions of the nuclei without concomitant divisions of the cytoplasm. This state, known as syncytium, is represented by one cell containing between 5000 and 6000 nuclei sharing the same cytoplasm. By the end of the eighth nuclear division, most nuclei have migrated to the surface, surrounding

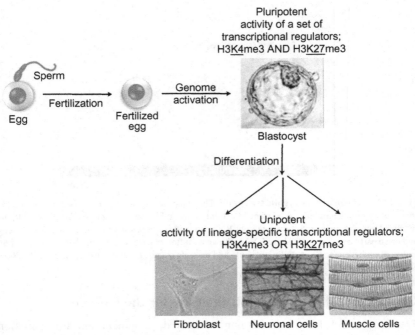

FIG. 13.4 Pluripotency and differentiation of embryonic cells. The transcriptionally inactive genome following fertilization is activated to give rise to cells in the blastocyst, which are pluripotent. These cells can give rise to all lineages of the developing organism. At the molecular level, these cells contain a number of transcriptional activators; on the other hand, transcriptional regulators that determine specific cell lineages are not expressed at significant levels in these pluripotent cells. These cells are often marked by bivalent chromatin domains that contain both activating and repressive histone modification marks (specific trimethylated lysine residues in histone H3, H3K4me3 and H3K27me3, respectively). These bivalent chromatin domains are later resolved during the process of differentiation into regions that contain only one type of these histone marks, trimethylated lysine K4 in transcriptionally active regions, and trimethylated lysine K27 in transcriptionally repressed regions. Additionally, during differentiation, lineage-specific gene regulators are activated. *(Based on Vastenhouw, N.L., Schier, A.F., 2012. Curr. Opin. Cell Biol. 24, 374–386, with permission. Cell images from Wikipedia, Creative Commons and https://socratic.org.)*

the yolk sac. After the 10th division, the pole cells form at the posterior end of the embryo, segregating the germ line from the syncytium. Finally, after the 13th division, cell membranes slowly invaginate, dividing the syncytium into individual somatic cells. The completion of this process marks the beginning of the gastrulation state of development.

Later stages of embryonic development lead to the formation of the adult fly body, which consists of 14 segments (Fig. 13.5). Three segments make up the head with its antennae and mouth parts. Three segments make up the thorax, with each thoracic segment having a pair of legs. In addition, the middle thoracic segment carries a single pair of wings. Finally, there are eight abdominal segments.

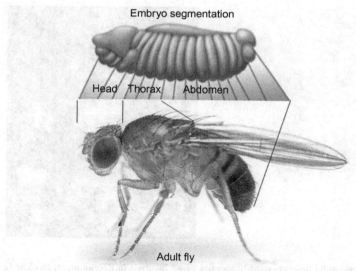

Embryo segmentation

Head Thorax Abdomen

Adult fly

FIG. 13.5   The segmented body of embryo and adult *Drosophila* fruit fly. See main text for description. (*Adapted from https://en.wikipedia.org/wiki/Evolutionary_developmental_biology and http://www.yourgenome.org.*)

The availability of *Drosophila* as a model organism led to significant progress in the study of the genetic control of development. Mutant forms of both the adult fly and the larvae allowed the identification of genes involved in developmental control. In the 1940s, Edward B. Lewis used chemical and radiation mutagens to isolate the famous four-winged mutant of the fruit fly; the wild-type fly has only two wings (Fig. 13.6A). He identified a cluster of genes (bithorax complex, BX-C) that regulate development of the insect's body segments. He eventually found that the linear arrangement of these genes on chromosomes was in an order corresponding to the fly's body parts; this order was later found to hold true for mice, humans, and other vertebrates. The concept is known as the colinearity principle.

Based on these studies, Lewis developed or reinforced, in the early 1960s, three key concepts about development, as succinctly stated by Duncan and Montgomery (2002): "(1) the BX-C genes are expressed locally within certain *Drosophila* body regions and are both necessary and sufficient for specifying their morphological identities; (2) The spatial expression of the BX-C genes is controlled by *cis*-regulatory elements that are defined by a specific class of mutations; and (3) these *cis*-regulatory elements could be pictured as responding to an anterior-posterior gradient of morphogens present early in development". Morphogens are defined as long-range **signaling molecules** that act over distances of a few to several dozen cell diameters. Morphogens diffuse through the tissues of an embryo, setting up concentration gradients. This graded distribution of morphogens within the embryo exposes cells to different morphogen

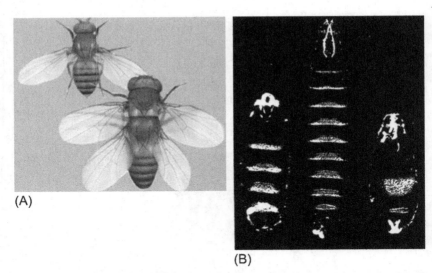

FIG. 13.6 Drosophila mutants used to identify genes involved in development. (A) The famous four-winged Drosophila mutant studied by Edward Lewis. Mutations in four genes give the fly in the lower right an extra pair of wings. (B) The cuticular pattern of segmentation mutants. A normal larva is shown in the middle, and two mutants on the left and right. The left mutant is characterized by approximately half the normal number of segments, every other segment has been deleted. In the right mutant most of the abdominal segments have been deleted. Anterior end of embryo is up. *(From © The Nobel Foundation, Nusslein-Volhgard.)*

concentrations, activating different transcriptional programs leading to different cells fates. In terms of their molecular structures, morphogens are either intracellular transcriptional regulators or extracellular growth factors that interact with extracellular matrix components.

A similar approach—mutagenesis, followed by identifying and studying mutant flies—was employed by Christiane Nüsslein-Volhard and Eric Wieschaus. This time the object of their research were fly larvae. The larva displays a clear axial organization with marks of position and polarity provided by the external cuticle, a tough but flexible outer protective layer for the developing larva. Using the cuticle for screening mutants was an ingenious choice at the time when there were no efficient methods for visualization of the hidden internal organs. In contrast, the skin and its cuticle were easily observed through simple fixation techniques, followed by microscopy. The mutants they identified had fewer bodily segments, gaps in the normal segment pattern, and alterations in other morphological characteristics of the segments (Fig. 13.6B). Later, identifying the genes in which mutation occurred led to defining a set of genes crucial for *Drosophila* embryogenesis. This, in turn, led to important new insights into the mechanisms that underlie the stepwise development of body segments.

The pioneering work of Edward Lewis, Christiane Nüsslein-Volhard, and Eric Wieschaus was recognized by the 1995 Nobel Prize in Physiology or Medicine "for their discoveries concerning the genetic control of early embryonic development." Importantly, these discoveries have broad significance for organisms other than fruit flies: many of the genes identified In *Drosophila* have homologues in other species, including humans.

INSIGHTS FROM A SIMPLE WORM

As described in Box 7.1, the nematode worm *Caenorhabditis elegans* has proved to be a powerful vehicle for the study of development at the cellular and molecular levels. Because the lineage of its 1090 somatic cells can be precisely mapped (see Figure 2 in Box 7.1), we can follow very specific developmental events from cell generation to cell generation. An excellent example is the development of the vulva. The worm is a hermaphrodite and fertilizes its own eggs internally, releasing them to the surroundings through a small orifice, the vulva. The development of this organ depends critically on the presence of one cell, called the "anchor cell" which attaches the gonad to the outer integument. The anchor cell recruits first three integument cells and eventually a total of 22 cells to form the vulva. If the anchor cell is specifically destroyed by laser radiation, the organ cannot form. Through mutation studies, it has been found that five genes are involved in this developmental step. Their protein products have been identified; one is essential for anchor cell function, the remainder for the surrounding vulvar cells. This work is approaching the ultimate in analysis of development, via a combination of genetics, cell biology, and biochemistry. With continued advances in single-cell techniques, this may anticipate the future of the field.

NUCLEAR TRANSFER EXPERIMENTS AND THE PRINCIPLE OF GENETIC EQUIVALENCE

The first experiments on this fascinating subject were based on nuclear transfer (Fig. 13.7). This method involves taking a cell nucleus from a donor and injecting it into an enucleated egg. This is then fertilized and allowed to develop as an embryo, which is then transplanted into the uterus of a foster mother. Success in such experiments began in 1952, when Robert Briggs and Thomas King cloned leopard frogs (*Rana pipiens*) by transferring nuclei from early embryo cells into enucleated eggs. Briggs and King cloned 27 tadpoles from 104 successful nuclear transfers. The first attempts to obtain normal tadpoles from nuclei extracted from differentiated cells failed. But in the 1960s, John Gurdon successfully cloned *Xenopus laevis* frogs starting from nuclei from intestinal epithelial cells of tadpoles, establishing once and for all the principle of genetic equivalence (see below). Sir John Gurdon shared the 2012 Nobel Prize for Physiology or Medicine with Shinya Yamanaka for "the discovery that mature cells can be reprogrammed to become pluripotent."

The experiments with amphibian cloning were fairly easy, because frog eggs are huge, a whopping 100,000 times larger than a normal somatic cell, allowing

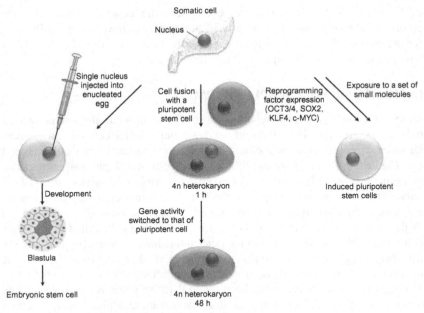

FIG. 13.7 **Different experimental strategies leading to pluripotency.** Reprogramming by somatic cell nuclear transfer: injecting a somatic cell nucleus into an enucleated oocyte can reset its memory of being a somatic cell; further development can generate an embryonic stem cell. Cell fusion-mediated reprogramming: fusing a somatic cell with a pluripotent stem cell erases the somatic cell identity. The resulting tetraploid cell (4n) exhibits pluripotent stem cell-like features. Transcription factor (TF)-mediated reprogramming: a small set of reprogramming factors transforms mature differentiated cells into induced pluripotent stem cells. Small molecule-mediated reprogramming: exposure to a set of small molecules also reprograms somatic cells to pluripotency. *(Adapted from Takahashi, K., Yamanaka, S., 2015. Development 142, 3274–3285, Fig. 3, with permission.)*

relatively easy micromanipulation. Working with the tiny eggs of mammals proved much more difficult. Nevertheless, over the next decade researchers were able to successfully clone a number of mammals, from mice to sheep. In 1997, the sheep "Dolly" was cloned from an epithelial cell of an adult donor, a major breakthrough that attracted public attention (see Chapter 17).

The cloning of animals (frogs, sheep, and other animals) from differentiated cells proved beyond doubt the so-called **principle of genetic equivalence**, which states that all cells of an adult organism contain the same genetic information and differ only in what portion of that information is being used in a specific cell type. Thus, nuclei from cells that are differentiated retain all the genetic information needed for the development of an entire organism; they just need to be "reprogrammed" by, for example, being transferred to the environment of an enucleated egg.

Interesting exceptions to this principle are the terminally differentiated antibody-producing cells. The genomes of such cells undergo considerable gene rearrangements to allow the production of the huge diversity of antibodies used by the immune system to fight disease agents, such as viruses and bacteria. Each cell and its progeny can encode just one type of antibody—termed **monoclonal antibody** because it is produced by a single cell clone—depending on the particular gene rearrangement that occurred in that cell. The normal immune response results in the production a large number of different antibodies, produced by clones of individual cells, each of which has undergone, during differentiation, a different kind of gene rearrangement. Interestingly, but not totally unexpected, when nuclei were transplanted from antibody-producing T-cells, the adult nuclear-transplant mice were normal but produced just only one kind of antibody. The gene rearrangement was irreversible!

GENOME REPROGRAMMING TOWARD EARLIER PHASES OF DEVELOPMENT IS POSSIBLE

Despite the enormous effort invested into investigating the molecular basis of such reprogramming of differentiated cell nuclei to pluripotency, we are far from understanding the process in its entirety. We know that nuclear transfer to eggs causes a major reprogramming of the great majority (>95%) of genes; the expression of differentiation-specific genes is discontinued, whereas genes that are normally expressed during early development are activated. Factors involved in these changes in gene expression are demethylation of DNA from promoter regions and posttranslational modifications of histones, the abundant nuclear proteins of chromatin fibers (see Chapter 12). The first morphologically recognized response of transplanted nuclei in eggs or oocytes is a huge increase in nuclear volume, as much as 30-fold. This volume increase is attributed to decompaction of chromatin fibers which is correlated with transcriptional activation. In addition, there is a significant exchange of proteins between the transplanted nucleus and the cytoplasm of the egg/oocyte. Notably, histones from the cytoplasm become highly concentrated into the transplanted nuclei. The importance of studies of reprogramming cannot be overstated, since the experimental reversal of development and differentiation provides a unique opportunity to understand the normal processes.

An alternative method or cellular reprogramming is based on cell fusion of normal somatic cells with a pluripotent stem cell (Fig. 13.7). In the **heterokaryon** (the cell containing two nuclei of different origin) formed initially, the nucleus coming from the differentiated cell switches its gene activity to that of the pluripotent cell.

A major breakthrough in understanding reprogramming came from the studies of Shinya Yamanaka and colleagues, some 40 years after the first nuclear transfer experiments of John Gurdon. These researchers worked with ESC isolated from early embryo cells and kept in culture as immature cells that can, using appropriate cell-culture conditions, give rise to a variety of different cell types. Yamanaka aimed at identifying the genes that make ESCs pluripotent.

He demonstrated that mouse and human embryonic or adult fibroblasts can be induced toward pluripotency by the induced expression of only four protein factors. These factors were chosen because some of them are known to function in the maintenance of pluripotency in both early embryos and ES cells. Others are known to be frequently upregulated in tumors and contribute to the rapid proliferation of ES in culture. The cells reprogrammed in this way were appropriately called **induced pluripotent stem cells** (**iPSC**). An alternative, more recent method of obtaining iPS cells from differentiated cells is through exposure of cultures to a set of small molecules (Fig. 13.7).

For the sake of completeness and fairness, we should mention much earlier experiments from Harold Weintraub and coworkers, who in the late 1980s did similar experiments in reverse. They were successful in activating muscle genes in a large variety of differentiated cells lines from chicken, rat, and human origin, by expression of an exogenous (inserted) copy of the gene *MyoD*, which is the master regulatory gene for muscle formation. The experiments demonstrated that no additional tissue-specific factors other than MyoD are needed to activate a slew of genes needed for terminal muscle differentiation.

In their totality, the genome reprogramming experiments, independently of the way reprogramming is achieved, pave the way to better understanding normal development, which still remains an enigma in terms of its molecular mechanisms. Such an understanding is critical for future use in the clinic.

EPILOGUE

As a new science develops, it often begins to spill over into and influence adjacent fields. That has happened here. For decades, for centuries, biochemistry, genetics, and embryology went their parallel ways, advancing but rarely interacting. Then the synthesis to biology at the molecular level suddenly provided new insights and techniques to approach developmental problems. In chapters to come, we will see this happening in other fields as well: botany, parasitology, paleontology, anthropology, to name a few.

FURTHER READING

Books and Reviews

Alberts, B., Bray, D., Lewis, J., Raff, M., Roberts, K., Watson, J.D., 1994. Molecular Biology of the Cell, third ed. Garland Publishing, New York, NY. The classic textbook in the field; chapter 22 is especially important to this topic.

Duncan, I., Montgomery, G., 2002. E. B. Lewis and the Bithorax Complex: part II. From cis-trans test to the genetic control of development. Genetics 161, 1–10. Part of a series of perspective articles "Perspectives: Anecdotal, Historical and Critical Commentaries on Genetics", edited by Crow, J. F. and Dove, W. F. Describes Ed Lewis's journey in science—from genetics, to molecular genetics, to the post-genomic era.

Evans, M., 2011. Discovering pluripotency: 30 years of mouse embryonic stem cells. Nat. Rev. Mol. Cell Biol. 12, 680–686. Historical perspective on the isolation of pluripotent cells from mice.

Gurdon, J.B., Byrne, J.A., 2003. The first half-century of nuclear transplantation. Proc. Natl. Acad. Sci. USA 100, 8048–8052. An overview of the amazing field of transplantation of nuclei that led to numerous insights into the molecular mechanisms of development and differentiation.

Hochedlinger, K., Plath, K., 2009. Epigenetic reprogramming and induced pluripotency. Development 136, 509–523. A comprehensive review of genome reprogramming with extensive definitions of terms and historic background.

Serafini, A., 2001. The Epic History of Biology. Perseus Publishing, Cambridge, MA. Good for detailed accounts of the earlier research.

Waddington, C., 1966. Principles of Development and Differentiation. Macmillan Company, New York, NY. One of numerous books by the British developmental biologist, paleontologist, geneticist, embryologist and philosopher that outlines his views on development and differentiation.

Yamanaka, S., 2012. Induced pluripotent stem cells: past, present, and future. Cell Stem Cell 10, 678–684. A personal account by the Nobel Prize winner about the development and use of the technique for creating pluripotent embryonic cells in culture by genetic manipulations.

Classic Research Papers

Bernstein, B.E., Mikkelsen, T.S., Xie, X., Kamal, M., Huebert, D.J., Cuff, J., Fry, B., Meissner, A., Wernig, M., Plath, K., Jaenisch, R., Wagschal, A., Feil, R., Schreiber, S.L., Lander, E.S., 2006. A bivalent chromatin structure marks key developmental genes in embryonic stem cells. Cell 125, 315–326. Genome-wide analysis revealing the existence of chromatin domains with specific combinations of histone modifications.

Briggs, R., King, T.J., 1952. Transplantation of living nuclei from blastula cells into enucleated frogs' eggs. Proc. Natl. Acad. Sci. USA 38, 455–463. The first successful transfer of nuclei from early embryo cells into enucleated eggs that led to the development of frog tadpoles.

Lewis, E.B., 1978. A gene complex controlling segmentation in Drosophila. Nature 276, 565–570. A report on the discovery, in adult mutant fruit flies, of a set of genes controlling development.

Martin, G.R., 1981. Isolation of a pluripotent cell line from early mouse embryos cultured in medium conditioned by teratocarcinoma stem cells. Proc. Natl. Acad. Sci. USA 78, 7634–7638. Describes the isolation of embryonic stem cells, an achievement of historic proportions.

Nüsslein-Volhard, C., Wieschaus, E., 1980. Mutations affecting segment number and polarity in Drosophila. Nature 287, 795–801. The original paper by the two Nobel Prize winners on identifying genes that control the segmentation pattern of Drosophila larvae.

Takahashi, K., Yamanaka, S., 2006. Induction of pluripotent stem cells from mouse embryonic and adult fibroblast cultures by defined factors. Cell 126, 663–676. The original report by Yamanaka's laboratory on the genetic manipulation of fibroblast cells grown in culture to revert them to pluripotency.

Chapter 14

Recombinant DNA: The Next Revolution

PROLOGUE

Knowledge leads to power. In the preceding chapters, we have followed the slow accumulation of knowledge about the basic mechanisms of life—how information is carried from generation to generation, and how it is expressed in the formation of the requisite molecular structures and processes. As the picture clarified, it did not take long for clever men and women to realize that this knowledge might lead to fantastic applications in sister sciences, in medicine and pharmacology, and in fields yet unimagined. In the following chapters, we shall explore some of those applications.

THE POWER OF DNA RECOMBINATION

The key to these applications was the introduction of **recombinant DNA technology**. These techniques, devised in the early 1970s, are very versatile. For example, they allow the isolation, copying, and amplification of any gene, thus generating DNA of the amount and purity necessary to determine the gene's nucleotide sequence. Alternatively, they provide a way to express specific genes at a high level so that their products, most importantly proteins, can be investigated or used medicinally. Finally, recombinant DNA technology makes it possible to introduce any desired mutation at a predetermined position in any gene, thus allowing the production of proteins with altered properties. Such mutated genes can be introduced into living cells as a way of studying their biochemical and physiological roles. In 1975, Sydney Brenner stated "Recombinant DNA technology will make a lot of things obsolete, the possible easier, the impossible possible...We can put duck and orange DNA together—with a probability of one...Probably more profound in the questions it can reach than any previous methods" (conversation of Brenner with Judson, 1979).

From a purely practical viewpoint, recombinant DNA technology has given rise to the contemporary biotechnology industry that produces hundreds of new drugs and vaccines. It also underpins the production of genetically modified crops with desirable characteristics. It has given rise to gene therapy projects that aim to cure genetic diseases. It has revolutionized scientific criminal

The Evolution of Molecular Biology. https://doi.org/10.1016/B978-0-12-812917-3.00014-0

investigations. Last, but not least, it forms the basis of efforts to protect biological species from extinction, and to create "hybrid" species that contain genes from long-extinct species as part of their present-day genomes. These are described in detail in a separate chapter.

Here, we introduce the benchmark techniques that collectively form the basis for the creation and use of recombinant DNA molecules. The development of such techniques took place in between 1960 and 1970. We begin with a general introduction to the concept and the individual steps of the fundamental technique in recombinant technology: the making of multiple copies of a DNA sequence by the technique called **cloning**.

HOW TO CLONE DNA

A **clone** is a set of identical copies of some biological entity; a bacterial colony grown from a single bacterium is a clone. Cloning of DNA is the production, using some **host organism** that can reproduce to a large number, of multiple copies of an inserted DNA sequence.

All cloning projects begin with two essential components: the piece of DNA to be cloned and the vector molecule that will be used to introduce the sequence of interest into a living host cell (Fig. 14.1). **Vectors** are DNA molecules that can incorporate the foreign nucleotide sequence, carry it into a cell, and then assure that it be maintained stably in that cell during repeated rounds of replication (Box 14.1).

The first step is the insertion of the donor DNA segments into the vector. This is simply a cut-and-paste process; it requires special enzymes, including restriction endonucleases for cutting and DNA ligases for resealing (see below). The resulting recombinant vectors are next introduced into living cells. There are a variety of ways to do this, each optimized for the specific host, be it a bacterium or a eukaryotic cell. Exposing host cells to the recombinant vector leads to several different types of products: those cells that did not take up any vector, those that contain the unaltered self-ligated vector, and those that contain recombinant vectors. Only a relatively small proportion of those cells with recombinant vectors will have the recombinant of interest. In order to fish the clone of interest from this heterogeneous cell population, various selection (or screening) methods are applied (Fig. 14.2).

CONSTRUCTION OF RECOMBINANT DNA MOLECULES NEEDS RESTRICTION ENDONUCLEASES AND LIGASES

Two classes of DNA enzymes are essential for creating any recombinant DNA molecule: restrictases to cut DNA and ligases to reseal it. The discovery of **restriction endonucleases** (also called **restrictases**) is one of the most important developments in molecular biology, for they are essential for all recombinant DNA technology.

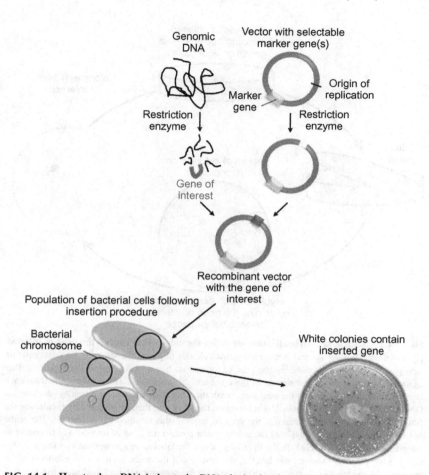

FIG. 14.1   How to clone DNA in bacteria. DNA cloning involves several steps. (1) Isolate DNA from an organism and prepare cloning vector with marker gene. (2) Produce recombinant DNA molecules via treatment of both donor DNA and vector DNA with a restriction endonuclease and a ligase. The restriction endonuclease will digest the organismal DNA into a large set of linear DNA fragments. (3) Ligate the DNA fragments into the vector linearized by the restrictase treatment. This step results in three different kinds of vectors: those that will carry the gene of interest, those that will have other DNA fragments from the original mixture inserted, and those that self-ligate (for simplicity only the first is depicted). (4) Introduce the population of vector molecules into bacterial cells, plate the cell population onto solid agar plates, and allow individual cells to generate colonies on the plate. Four different kinds of bacterial cells are obtained following the vector insertion procedure, reflecting the three different kinds of vectors described earlier, and one that has not taken up the vector. Finally, screen for clones that contain the gene of interest. The incorporation of genes for antibiotic resistance in the vector plasmid allows elimination of cells that did not take up the vector and those that have self-ligated vectors. Finding the bacterial clone that carries the gene of interest against the background of all other recombinant bacteria is done by colony hybridization with a labeled probe that contains the gene of interest (see Fig. 14.2).

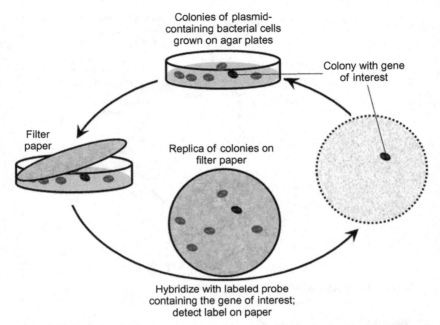

Colonies of plasmid-containing bacterial cells grown on agar plates

Colony with gene of interest

Filter paper

Replica of colonies on filter paper

Hybridize with labeled probe containing the gene of interest; detect label on paper

FIG. 14.2 **Nucleic acid hybridization identifies the colony that carries the gene of interest.** Hybridization involves several steps. First, grow colonies of plasmid-containing bacterial cells on agar plates. The colony containing the gene of interest is shown in *dark gray*. Then, press a filter against colonies to produce a replica of the colonies; this is possible since some cells from each colony adhere to the filter. In the next step, wash the filter with an alkali solution to denature the DNA. The resulting ssDNA, which is retained on the filter, can then be detected by incubating the filter with ssDNA (probe) containing the gene of interest (this is called hybridization). The probe has been labeled in some way; only the colonies that contain the gene of interest will hybridize to the probe and become labeled. In the final step, detect the label in an appropriate way (autoradiography if the probe is radioactively labeled; fluorescence if the probe carries a fluorophore). Then compare the pattern of colonies with the probe with the pattern on the original plate to identify a colony containing the gene.

BOX 14.1   Cloning Vectors

A very large repertoire of cloning vectors based on plasmids, viruses, or combinations thereof are used in recombinant DNA technology. Which one would choose depends on the target cell, the amount of DNA to be incorporated, and the purpose of incorporation.

Plasmids are extrachromosomal nucleic acid molecules carried by bacteria that replicate independently of the main bacterial chromosome. Plasmids are generally dispensable for the host cell but may carry genes that are beneficial under certain circumstances. The expression of most of these genes confers antibiotic resistance or is responsible for antibiotic production and can thus be used for selection of bacteria into which plasmids have been introduced.

BOX 14.1   Cloning Vectors—Cont'd

The number of derivative plasmids that have been used for cloning is enormous. Figure 1 presents one classical example of plasmid cloning vectors.

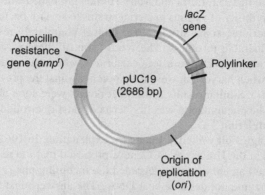

FIGURE 1.   pUC19, an example of a classical bacterial cloning vector. pUC19 (UC stands for University of California, where the vector was constructed by Joachim Messing and colleagues) contains, in addition to the standard antibiotic selectable marker and origin of replication, a gene that encodes the enzyme β-galactosidase (lacZ). A polylinker (an array of multiple restriction sites for cloning, also known as multiple cloning site, MCS), splits the *lacZ* gene. Thus, transformed cells containing the plasmid with inserted gene can be distinguished from cells containing the original (nonrecombinant) plasmid by the color of the colonies they produce: recombinant cells will be white, whereas the nonrecombinants will turn blue because the presence of active β-galactosidase will transform a colorless substrate present in the medium to a blue compound.

Plasmid vectors are often used for cloning small DNA segments (of up to 10 kbp). Cloning larger fragments requires the use of alterative vectors, such as derivatives of bacteriophages. Vectors that combine features from plasmids and phages—called cosmids and phagemids—have also been constructed and are in wide use. Finally, **artificial chromosomes** constructed in the laboratory meet the need for cloning very large DNA fragments in bacteria or yeast.

The Expression of Recombinant Genes

If the aim of a project is to produce large amounts of protein, whether native or mutated, specialized vectors called **expression vectors** must be used. They must ensure that the exogenous gene they carry will be transcribed, and in a form that can be translated into protein, using the host cell's machinery. Ideally, expression vectors should be able to integrate into the host genome to ensure stable propagation of the gene of interest from cell generation to cell generation. They should also carry a **promoter** that is functional within the given host (promoters are elements where RNA polymerases bind to initiate gene transcription). Finally, expression should be highly regulated, that is, the gene should be expressed only when desired. This is especially important if the protein happens to be toxic to the host cell.

The existence of enzymes that cleave nucleic acids was recognized as early as 1903, but the fact that some could make specific cleavages was realized only half a century later. The first hint of such activity came from seemingly unrelated genetic studies in the laboratory of the pioneering molecular geneticist, Salvatore Luria. In 1952, Luria and Mary Human published studies on the comparative susceptibility of various bacterial strains to several bacteriophages. The results were curious: some strains were almost completely resistant to some phage, but vulnerable to others. The word "almost" here is significant; even in resistant strains, there were a few bacteria in which the phage could grow; these phages, when harvested, were fully potent against the previously resistant strain of bacteria. Similar results, with other phage, were soon obtained in other laboratories. The phenomenon was first termed "host-controlled variation," and later "**host restriction**."

The puzzling result waited a decade for explanation. In 1962, Werner Arber and associates at the University of Geneva proposed that the resistant bacterial strains contained an enzyme that degraded the incoming phage DNA. But why did such an enzyme not degrade host DNA? The answer must be that the host DNA had somehow been modified by another enzyme to resist this degradation. Attachment of methyl ($-CH_3$) groups to DNA (methylation) by a methylase was suspected as the source of host resistance by Arber as early as 1965, but not experimentally demonstrated until 1972. The postulate of a "protecting" enzyme also explained the occasional observation of acquired infectivity: in rare cases, the DNA of invading phage could itself be methylated and thus protected from degradation. The importance of these findings to the development of biotechnology is an outstanding example of the unexpected consequences of "pure" or unapplied research.

Early studies on the restriction process centered largely on type I restriction endonucleases, which do not cleave at a specific site, and cannot be very useful in the laboratory. The major breakthrough came in 1970, when Hamilton Smith and colleagues at Johns Hopkins University discovered type II restriction endonucleases, which both recognize and cleave the DNA at a sequence-specific site. A year later, Kathleen Danna and Daniel Nathans, also at Johns Hopkins, used such an enzyme to digest the DNA of the virus SV40 into discrete fragments, which they separated by gel electrophoresis. Such separation of restriction fragments allowed for routine isolation and cloning of specific fragments of DNA. The enormous potential of this work was recognized by awarding, in 1978, the Nobel Prize in Physiology or Medicine to Arber, Smith, and Nathans "for the discovery of restriction enzymes and their application to problems of molecular genetics."

More than 3000 restrictases have been described to date. The name of each restrictase carries information about the bacterial species and strain in which it was originally discovered; a Roman numeral is used to differentiate between several enzymes from the same strain. For example, *Eco*RI designates one of the two known restrictases in *Escherichia coli* strain R; *Hae*III denotes the third enzyme present in *Haemophilus aegyptius*.

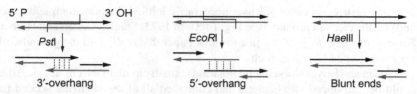

FIG. 14.3 **The cleavage action of type II restriction endonucleases.** Type II restriction enzymes cleave the polynucleotide DNA chain internally, to produce 5'-phosphates and 3'-OH groups. Cleavage sites on the two DNA strands can be offset by as many as four (or more) nucleotides to produce overhangs (also known as sticky ends); when there is no offset of the cutting positions, blunt-end products are created. Specific examples of such cleavages are presented.

Each restriction enzyme is characterized mainly by the specific sequence it recognizes and the site where it cuts the double helix. Different enzymes can cut the opposite strands of the DNA helix, either in a staggered way, producing **overhangs** (also known as **sticky ends**), or exactly opposite of each other, producing **blunt** (or **flush**) **ends** (Fig. 14.3). Both kinds of products can be ligated to form uninterrupted molecules, but the ligation efficiency of blunt-ended fragments is much lower than that of fragments containing overhangs. This is because the sticky ends can actually base-pair with each other, increasing the probability of the two ends staying in close proximity and in proper orientation for ligation to occur.

DNA ligases are indispensable for numerous cellular processes, including DNA replication, recombination, and repair, whenever there is a need to seal nicks in the DNA double helix or join broken helices. Ligases form phosphodiester bonds between the 3'-OH and 5'-phosphate termini of the combining DNA fragments.

THE FIRST RECOMBINANT DNA MOLECULES

The availability of site-specific endonucleases and DNA ligases made it possible to create the first recombinant DNA molecules in the early 1970s. In 1972, Paul Berg and his group at Stanford University carried out the first experiments to specifically modify a natural DNA. They used a restrictase from *E. coli* to cleave the DNA chromosome of the mammalian virus SV40 at one specific point, opening the closed circle into a linear form. Some additional steps allowed for the construction of dimers and higher oligomers of SV40 DNA, and for the insertion of foreign DNA fragments into the viral DNA.

This was a powerful and ground-breaking result: it was the first time that a natural DNA molecule had been specifically modified in a premeditated way. In 1980, Paul Berg was awarded the Nobel Prize in Chemistry for "his fundamental studies of the biochemistry of nucleic acids, with particular regard to recombinant-DNA." The technique was, however, somewhat awkward in that it required additional steps before the broken molecules could be ligated together. Shortly thereafter, it was recognized that the very nature of the cut made by many restriction endonucleases necessarily left overlapping complementary ends on the cleaved product (see Fig. 14.3). In 1973, Stanley Cohen and Herbert Boyer took advantage of this in a seminal paper that really laid the groundwork for recombinant DNA research.

Cohen and Boyer worked with **plasmids** (small circular DNA molecules) that could be introduced into bacteria and contained all of the elements needed for their replication in the bacterial cell. They also contained expressible genes for resistance to various antibiotics, which allowed rapid and efficient screening for colonies of bacteria containing each plasmid. Finally, each contained an *Eco*RI site, placed so as not to interfere with plasmid replication or gene expression. Cohen and Boyer were able, by taking advantage of the self-complementarity of the DNA ends produced by *Eco*RI cleavage, to make recombinant combinations of antibiotic resistant plasmids, and then "clone" these in the bacterial host. Then, they used some of these recombinant plasmids to clone the amphibian rRNA gene (Fig. 14.4).

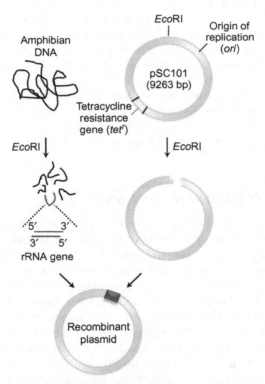

FIG. 14.4 The experiment of Cohen and Boyer to produce recombinant DNA molecules carrying the rRNA gene from an amphibian. The resulting recombinant plasmids were capable of replicating in *E. coli* and could be selected for because the antibiotic-resistant gene they carried was active. This allowed bacteria transformed with the recombinant plasmid to live and propagate in antibiotic-containing medium whereas nontransformed bacteria died out.

Cohen and Boyer did not miss the much broader implications of their work. In the concluding paragraph of their 1973 paper, they stated: "The general procedure described here is potentially useful for insertion of specific sequences from prokaryotic or eukaryotic chromosomes or extrachromosomal DNA into independently replicating bacterial plasmids." Considering the enormous range that recombinant DNA methods have since enveloped, the statement is remarkably modest.

In 1977, Boyer's laboratory and outside collaborators described the first-ever synthesis and expression of a peptide-coding gene (somatostatin). In August 1978, they produced synthetic insulin, followed in 1979 by a growth hormone. Cohen and Boyer have won numerous prestigious awards for their major contributions to both basic research and the practical applications of recombinant DNA technology.

POLYMERASE CHAIN REACTION AND SITE-DIRECTED MUTAGENESIS

In the late 1970s and early 1980s, two methods were introduced that became indispensable for the further development of recombinant DNA technology. Both methods won Nobel Prizes soon after they were introduced, in recognition of their immediate and broad impact. The first method, **polymerase chain reaction** (**PCR**), makes it possible to prepare substantial quantities of specific DNA fragments from minute amounts of starting material. This in vitro method uses repetitive cycles of denaturation, polynucleotide synthesis, and renaturation (annealing) (Fig. 14.5). Synthesis proceeds from small synthetic primers complementary to the ends of the sequence of interest. PCR has proved of enormous value in extracting sequence information from minute amounts of DNA in fossils. It has been indispensable in modern crime detection, allowing DNA profiling (see Chapter 17) from as little material as a single hair. PCR is also indispensable for the analysis of complex natural populations of microorganisms, most of which cannot be grown under laboratory conditions.

The second method, of enormous scientific and medical importance, is **site-directed mutagenesis**. For many years, scientists have been on the hunt for methods to mutate DNA sequences, mainly to study the effect of mutated genes, and hence their altered protein products, on protein properties and function. The initial approaches used chemicals to modify an amino acid in the polypeptide chain directly (without modifying the gene sequence), or used physical or chemical agents to mutagenize the DNA coding for the protein in vivo. Both methods suffered from a major drawback, they were unable to change one specific amino acid at one predetermined position within the molecule. Both methods produced complex sets of molecules that had to be further fractionated and characterized in order to isolate the molecule of interest.

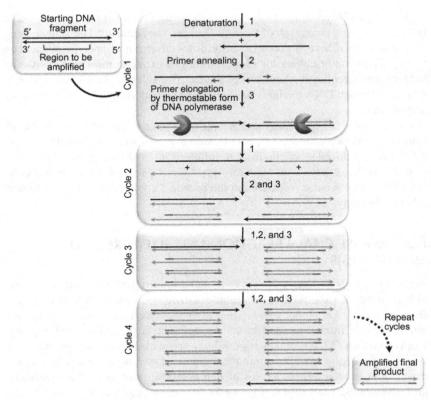

FIG. 14.5   The PCR amplification reaction. The reaction usually consists of a series of 20–40 repeated cycles, where each cycle consists of three discrete temperature steps: denaturation, annealing, and elongation. During the denaturation step, the DNA sample is heated to 94–98°C for 20–30 s: the high temperature disrupts the H bonds between complementary bases and separates the two DNA strands. During the annealing step, the temperature is lowered to 50–65°C for 20–40 s to allow annealing of the primers (15–30-nucleotide-long synthetic DNA fragments) to the single-stranded DNA template. The primers are complementary in sequence to the ends of the region of interest and possess a free 3'-OH group which is needed for the action of DNA polymerase at the third step.

A major breakthrough occurred in 1978, when Michael Smith and colleagues at the University of British Columbia in Vancouver, Canada, came up with a relatively simple in vitro procedure to create any mutation at any preselected site. The procedure underwent many modifications, but the principle, illustrated in Fig. 14.6 stays the same.

Michael Smith won the 1993 Nobel Prize in Chemistry for "his fundamental contributions to the establishment of oligonucleotide-based, site-directed mutagenesis and its development for protein studies." Smith understood what it took to be successful in science: "In research you really have to love and be committed to your work because things have more of a chance of going wrong than right. But when things go right, there is nothing more exciting." The prize was shared with Kary Mullis for "for his invention of the polymerase chain reaction (PCR) method."

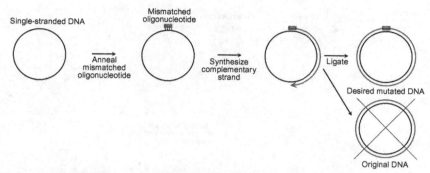

FIG. 14.6 Simplified schematic illustrating the principle of site-directed mutagenesis. The single-stranded DNA could be that of the single-strand DNA phage M13 containing the cloned gene; alternatively, if the gene had been cloned in a double-strand vector, the DNA needs to be first denatured, to allow hybridization with the synthetic oligonucleotide. Several rounds of replication of the ligated product introduced into a bacterial cell will result in the formation of a mutated sequence of the original DNA. Convenient methods exist to get rid of the latter. *(From Molecular Biology: Structure and Dynamics of Genomes and Proteomes by Jordanka Zlatanova and Kensal E. van Holde. Reproduced by permission of Taylor and Francis Group, LLC, a division of Informa plc. Copyright 2015.)*

MANIPULATING THE GENETIC CONTENT OF EUKARYOTIC ORGANISMS

The development of recombinant DNA techniques has given us the power to change, at will and in specific ways, the genomes of plants and animals to produce what are referred to as **transgenic organisms**. We shall discuss some of the many practical (and unpractical) applications of such techniques in Chapter 17; here we shall describe briefly how it is done.

In the early 1980s, a number of laboratories developed a technique for introducing new genetic material into animal genomes in a way that would be transmitted from generation to generation. An egg is removed from a female animal (e.g., a mouse) and fertilized in vitro. The vector containing the gene of interest and the nucleotide sequences necessary for its proper transcription is injected into the male pronucleus, that is, the haploid nucleus of the sperm, after the sperm enters the egg. The male pronucleus then fuses with the pronucleus of the egg to form the diploid nucleus of the embryo. The timing of the microinjection is critical because the introduced DNA must integrate into the genome prior to the duplication of the genetic material that takes place before the first embryo cell division. If integration occurs following that division, the mouse will be mosaic, with some wild-type and some transgenic cells. At the next stage, the egg is transplanted to the uterus of a "foster" mouse, and a pup, which carries the desired genetic change, will be born.

Site-directed mutagenesis makes it possible, in principle, to insert a protein with a new or modified sequence into an organism, and to study its effect on the whole organism. However, this is the addition of a function; the original variant of

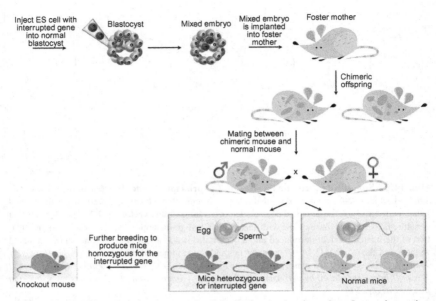

FIG. 14.7 Making a knockout mouse by introducing an in vitro altered gene into mice. *(Adapted from the website of the Nobel Foundation. Copyright: the Nobel Committee for Physiology or Medicine, 2007.)*

the gene is still present, and in most cases functional. If we want to take away a gene, or replace it with an altered version, the challenge is somewhat more difficult.

To inactivate, replace, or otherwise modify a particular gene, the vector must be "targeted" for **homologous recombination** at that particular site. This goal, long thought impossible, was achieved by the Italian-born American scientist Mario Capecchi in the late 1980s. Fig. 14.7 illustrates the widely used procedure to produce "**knockout**" mice, in which a specific gene has been inactivated by interrupting it with an extraneous sequence.

Modifications of these procedures can be used to generate "**knockin**" **organisms**, in which a modified gene is specifically substituted for the wild-type gene, or "**knockdowns**," in which the regulation of a particular gene is modified. All of these techniques are very time-consuming and relatively inefficient but can provide the most definitive evidence for the in vivo function of a particular gene. Capecchi shared the 2007 Nobel Prize in Physiology or Medicine with Sir Martin Evans and Oliver Smithies for their "discoveries of principles for introducing specific gene modifications in mice by the use of embryonic stem cells."

CRISPR, THE GENE-EDITING TECHNOLOGY OF TODAY AND TOMORROW

The introduction of this new technique in 2012/2013 by groups at the University of California at San Francisco and at Berkley is viewed by many as the "biggest game changer to hit biology since PCR" (Ledford, 2015). CRISPR-based systems can edit genes in living organisms in any desired way—adding or deleting gene sequences or changing sequences in a predetermined way. The technique is fast, extremely low cost ($30!), and easy to perform without any special training. It has a very wide range of applications, including gene therapy, creating plants and animals of desired characteristics, and eradicating pathogens, to name a few. What is this miracle tool and how did it come about?

The technique is based on a 1987 discovery by a Japanese group of the existence of a system in *E. coli* that protects against viral infection. Similar systems were later found in other bacteria and in *Archaea*. The genomic structure of the system, called **CRISPR-Cas** for **clusters of regularly interspaced palindromic repeats and associated proteins**, is shown in Fig. 14.8A. It contains intermediate-size repeats that are separated by spacer elements of similar length, some of which exactly match sequences from phages and plasmids, suggesting that these spacers were acquired during previous viral attacks on the

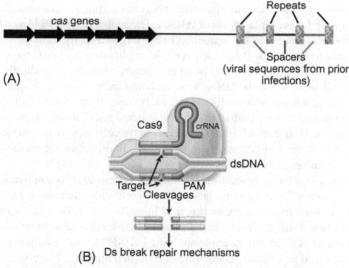

(A)

(B)

FIG. 14.8 The CRISPR-Cas system. (A) Simplified diagram of a CRISPR locus. The two major components of a CRISPR locus are shown: cas genes and a repeat-spacer array. Repeats are shown as grey boxes and spacers, the remnants of viral sequences from a prior infection as thicker lines bars. Several CRISPRs with similar sequences can be present in a single genome, only one of which is associated with cas genes. The viral sequences in an array act as "guides" to direct the associated cas9 protein to cleave the DNA at the selected site. (B) Cas9 nuclease site specifically cleaves double-stranded DNA activating double-strand break repair machinery. The end product of the repair process will depend on the specific type of repair system used by the cell. In addition, precise mutations and knock-ins can be made by providing a homologous repair template and exploiting the homology-directed repair pathway. (*(A) Adapted from Wikipedia. (B) Adapted from New England Biolabs web site 2017, https://www.neb.com/tools-and-resources/feature-articles/crispr-cas9-and-targeted-genome-editing-a-new-era-in-molecular-biology.*)

bacterium. The system also contains repeated *cas* (CRISPR-associated) genes, coding for a diverse group of proteins. Cas proteins are nucleases that cut DNA in both strands, thus creating a double-strand break.

CRISPR-Cas protects bacteria by destroying invading viral nucleic acid. The system acquires short DNA elements form invading phages and stores them in CRISPR arrays in the host genome. Upon reinfection, the spacer elements and the palindromic DNA are transcribed into long RNA molecules, which are then cut into smaller CRISPR RNAs (crRNAs) by the action of the cas nucleases. crRNA then guide surveillance complexes to complementary sequences in invading nucleic acids (Fig. 14.8B). The same cas nucleases that are involved in the production of crRNAs then cleave the invading DNA. Overall, what has happened is this: a previous encounter with a particular kind of bacteriophage has led to the production of RNA repeats containing a segment of the phage sequence. The transcript of the phage sequence forms a complex with a DNA-cleaving nuclease and will guide it to cut the region in the host genome DNA that is complementary to the transcript.

The system developed from this principle and used for gene-editing in a host cell is a recombinant plasmid that combines genes coding for nuclease Cas9 and genes coding for multiple copies of crRNA. The sequence of crRNA is designed to be complementary to the sequence being targeted in the host genome and thus needs to be designed for each specific application. In brief, the CRISPR technique allows us to select *any* point in a genome and design a simple system that will allow us to cut the DNA there, and only there. Once such specific cuts can be made, it is possible to disable genes, remove them, or substitute other or modified genes. Since the change is in the genome itself, it will be passed on to the descendants of that cell *in perpetuity*. If germ cells (sperm or ova) are used, a whole new variety of organism can be created. In principle, the method can be applied to human cells as easily as any other.

Research on the gene-editing technology system and its applications has become a major focus in many laboratories worldwide. Because it is so cheap and easy to perform, there are no practical limits to its use. Despite the huge potential there are still environmental and ethical issues that need to be addressed. The foremost of these is concern of applying CRISPR to viable human embryos. This could lead to a stable genetic line of modified humans! The concern of the international community reached a peak when a group of Chinese researchers reported in April of 2015 that they had successfully used CRISPR-Cas to cleave the endogenous β-globin gene in nonviable human zygotes. The application of the technique has now been approved by the National Institutes of Health for only somatic cells, but not the germline in embryos.

EPILOGUE

Molecular biology originated from a fusion of biochemistry and genetics that proved capable of explaining what genes really are, how they carry information from generation to generation, and how that information is expressed. Along the way, techniques were developed that greatly increased our power to "read" the genome, or to modify it in specific ways. These, in turn, yielded remarkable practical results and procedures, in other sciences, medicine, and biotech industries. To these we turn in the following, concluding chapters.

FURTHER READING

Books and Reviews

Capecchi, M.R., 2005. Gene targeting in mice: functional analysis of the mammalian genome for the twenty-first century. Nat. Rev. Genet. 6, 507–512. An overview of gene targeting by a Nobel Prize winning researcher.

Cohen, S.N., 2013. DNA cloning: A personal view after 40 years. Proc. Natl. Acad. Sci. USA 110, 15521–15529. A personal report by a Nobel Prize winner about the research that led to the development and broad applications of recombinant DNA technology.

Glick, B.R., Pasternak, J.J., Patten, C.L., 2010. Molecular Biotechnology: Principles and Applications of Recombinant DNA, fourth ed. ASM Press, Washington, DC. A recent book covering the major principles behind recombinant DNA technology and its applications.

Glick, B.R., 2013. Medical Biotechnology, first ed. ASM Press, Washington, DC. A groundbreaking textbook illustrating the medical applications of biotechnology.

Kunkel, T.A., Roberts, J.D., Zakour, R.A., 1989. Rapid and efficient site-specific mutagenesis without phenotypic selection. In: Wu, R., Grossman, L., Moldave, K. (Eds.), Recombinant DNA Methodology. Academic Press, San Diego, CA, pp. 587–601.

Nathans, D., Smith, H.O., 1975. Restriction endonucleases in the analysis and restructuring of DNA molecules. Annu. Rev. Biochem. 44, 273–293. An excellent early review on restriction endonucleases by two Nobel Prize winners, discoverers of the enzymes.

Ledford, H., 2015. CRISPR, the disruptor. Nature 522, 20–24. An overview of how the gene-editing technology emerged and evolved. Discusses ethical issues and potential future applications.

Mullis, K.B., 1990. The unusual origin of the polymerase chain reaction. Sci. Am. 262, 56–65. The Nobel Prize winner describes the unusual circumstances, "an improbable combination of coincidences, naiveté and lucky mistakes" that led to the development of the polymerase chain reaction.

Classic Research Papers

Cohen, S.N., Chang, A.C., Boyer, H.W., Helling, R.B., 1973. Construction of biologically functional bacterial plasmids *in vitro*. Proc. Natl. Acad. Sci. USA 70, 3240–3244. The seminal paper that laid the groundwork for recombinant DNA research.

Hutchison III, C.A., Phillips, S., Edgell, M.H., Gillam, S., Jahnke, P., Smith, M., 1978. Mutagenesis at a specific position in a DNA sequence. J. Biol. Chem. 253, 6551–6560. The paper sets the stage for the development of the oligonucleotide-based, site-directed mutagenesis.

Jackson, D.A., Symons, R.H., Berg, P., 1972. Biochemical method for inserting new genetic information into DNA of Simian Virus 40: circular SV40 DNA molecules containing lambda phage genes and the galactose operon of *Escherichia coli*. Proc. Natl. Acad. Sci. USA 69, 2904–2909. A ground-breaking paper that describes the modification of a natural DNA molecule in a premeditated way.

Chapter 15

Understanding Whole Genomes: Creating New Paradigms

PROLOGUE

The molecular approach has revolutionized almost all aspects of biology, as well as related sciences like evolution and anthropology. This has largely depended on the development of methods that allow the determination of whole-genome sequences. Such tasks, at least for higher organisms, were thought impossible a few decades ago; today they are routine. In this chapter, we will briefly trace the evolution of the methodology that has made this possible and then explore a few of the remarkable, often unexpected results.

THE EVOLUTION OF SEQUENCING METHODOLOGY

Fred Sanger, who had already become famous for the invention of protein sequencing methods (Chapter 3), decided, in the early 1970s, to turn his attention to nucleic acids. In studying the copying of single-strand DNA in the presence of radioactive dATP, he noted a curious result. At that time, radiolabeled nucleotides were in short supply, so Sanger had reduced the amount of dATP, compared to other deoxynucleotides. Surprisingly, the consequence was that a lot of the synthesized chains were much shorter than expected. He guessed correctly that this was because chains were apt to terminate just before a scarce A was encountered, and that this phenomenon could be extended to other bases. By that time, gel electrophoresis had been developed to single-base resolution. Thus, such gels, prepared from four different polymerization reactions, each deficient in one building block, could be used to "read" the sequence. Other, somewhat different methods were soon developed, but Sanger's basic observation, facilitated by vast advances in automated technology, has provided the basis for most methods utilized to this day.

The improvement has been spectacular. In the early days, sequencing a piece of DNA 100 base pairs in length would require days or weeks. Now, as we will describe below, "superfast" automated methods can handle gigabases of DNA in hours. The development of methods can be tracked by the corresponding dates of presentation of some whole-genome sequences. The size of genome that can be sequenced has increased a million-fold since 1977. We now have whole-genome sequences from every branch of the tree of life.

The Evolution of Molecular Biology. https://doi.org/10.1016/B978-0-12-812917-3.00015-2

GENOMIC LIBRARIES CONTAIN THE ENTIRE GENOME OF AN ORGANISM AS A COLLECTION OF RECOMBINANT DNA MOLECULES

Until very recently, it was not possible to directly sequence long stretches of DNA. However, soon after the success in cloning and sequencing of individual genes (Chapter 14) scientists realized that it would be possible, using recombinant technology, to prepare large sets of recombinant DNA molecules—vectors plus inserts—that can contain entire genomes in the form of overlapping sequences. If a sufficient number of overlapping sequences are known, these can be matched to deduce the whole sequence. For any but the very smallest genomes, computer analysis is necessary. These so-called **genomic libraries** can be multiplied in bacteria and stored indefinitely. The availability of such libraries was the most important prerequisite for the sequencing of whole genomes, first for viruses, then bacteria and yeast, then higher eukaryotes and culminating, in 2001, with the complete human genome (see below).

Creating complete genomic libraries can often be technically challenging. The size of the library needed to guarantee representation of all sequences of a genome is a function of the size of DNA fragments that can be accommodated in the cloning vector, the size of the genome itself, and the precision demanded. Thus, for example, the number of recombinant clones in a human library should be close to 700,000 to include any desired sequence with a probability of 99%.

Genomic libraries are not the only type of libraries important to research and biotechnology. Very useful libraries have been created to represent the messenger RNA (mRNA) contents of eukaryotic cells. Each cell type in an adult eukaryotic organism is characterized by a different set of mRNA molecules; some of these are shared among all cells and some are unique to that specific cell type. The unique mRNA molecules are the ones that are responsible for the differentiated phenotype of that cell type—they code for proteins that perform highly specialized functions.

The best way to study which protein genes are transcribed—and later translated—in a given cell type is to isolate the population of mature mRNA from the cytoplasm, and then, using in vitro methods (reverse transcription), to produce double-stranded DNA copies of this population. These are known as **cDNAs** (for copy DNAs). Cloning this entire population of cDNAs into bacterial vectors produces **cDNA libraries**. These can be maintained stably in bacteria and stored indefinitely, exactly as genomic libraries are. The existence of such libraries has practical applications: because they lack introns, cDNA clones are a convenient source of sequences to be directly expressed in bacterial cells.

THERE ARE TWO CLASSIC APPROACHES FOR SEQUENCING LARGE GENOMES

In the **whole-genome shotgun approach**, the entire genome is cloned as a series of recombinants, and each clone is sequenced. Computational methods are then used to reconstruct the entire genome, through alignment of clones that contain overlapping sequences. Such overlapping clones always exist because the initial random fragmentation of the genome is performed on a population of cells, each of which will produce its own random set of fragments. The second strategy for genome sequencing is the **hierarchical shotgun sequencing approach** (also called BAC-based or clone-by-clone sequencing). This approach involves generating and organizing (through restriction nuclease mapping) a set of large insert clones, typically 100–200 kbp each, and then separately performing shotgun sequencing on the appropriately chosen clones. The latter approach has been selected, following "lively scientific debate" (Lander et al., 2001), for sequencing the human genome.

ULTRAFAST SEQUENCING ALLOWS DEEP ANALYSIS OF GENOMES

The development of high-throughput, fast, and inexpensive methods for next-generation sequencing (NGS) occurs at an amazing pace, ushering in a new era of molecular biology research and, equally importantly, of data-based personalized medicine.

New techniques and significant improvements of old ones appear each year; these technical developments are supported by both federal government-funded institutes and by specialized private companies. These techniques are too many and too technical to describe here. It suffices to say that now more than 100 million bases can be read in just a few hours in an automated and very inexpensive way. The cost is less than $1.00 per 1 million bases. As an example, we mention the sequencer developed and marketed by 454 Life Sciences Corp.: one person can do the work of a hundred people and millions of dollars' worth of robots, sequencing an entire genome in a few hours at a very reasonable price.

For clinical purposes, NGS has practically replaced its predecessor, Sanger sequencing. Questions, though, are still lingering. Is NGS sensitive enough to catch all disease-associated sequence variants, at the same time avoiding false-positives? A recent (2016) large study looked at 20,000 samples from cancer patients and reported missing 2.2% of clinically relevant mutations. The Sanger confirmation of the NGS data cost $1.9 million, so it is still questionable how many of the companies performing these tests are willing to pay this additional cost. In addition, Sanger confirmation slows down the entire analysis.

WHOLE GENOMES

The availability of NGS allows sequencing of whole genomes from a variety of organisms. We now have numerous fully sequenced genomes from representatives of every branch of the tree of life—viruses, bacteria, unicellular eukaryotes, fungi, animals, and plants.

What kinds of questions can be asked and what insights do we get from these genomes? We will illustrate these issues on four recently (Fall, 2016)

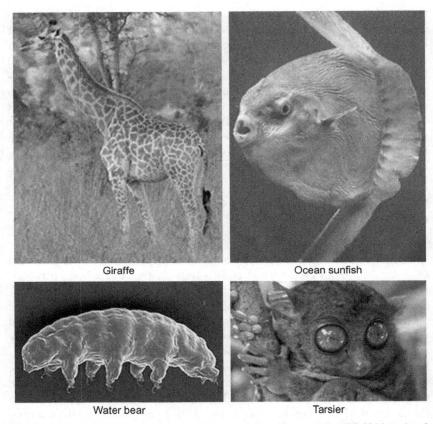

Giraffe   Ocean sunfish

Water bear   Tarsier

FIG. 15.1 Images of the animals whose entire genomes have been recently (Fall, 2016) analyzed.

published genomes (for images of the animals analyzed, see Fig. 15.1). The giraffe (*Giraffa camelopardalis*) genomes showed that there are at least four distinct species of giraffes. Researchers sequenced and compared seven loci in the genomes of 141 giraffes and were extremely surprised by the differences they found because the morphological and coat pattern differences between giraffes are limited. The newly acquired genome knowledge is expected to help in efforts to protect this species from extinction.

The ocean sunfish (*Mola mola*) attracted attention mainly by its very fast growing rate—nearly 1 kg a day—the adult reaches up to 2.7 m and 2.2 tons! It was found that numerous genes evolve faster than the corresponding genes in its relatives; many of these encode growth hormones.

The tardigrade (*Ramazzottius varieornatus*), also known as water bear, is a weird-looking, water-dwelling, eight-legged, segmented tiny animal, known to be able to withstand extreme environmental conditions: heat, cold, drought, high pressure, X-ray exposure. Probably not surprising, amplified levels of stress-tolerance genes were detected. For the first time, a DNA-associating protein was discovered, which confers DNA protection and improved tolerance to radioactivity. Furthermore, inserting this protein into cultured human cells induced the same X-ray protection, suggesting possible future use in space flights, radiotherapy, and for radiation workers.

Tarsiers (*Tarsius syricht*) are tiny primates with large golden eyes, long fingers, and squat furry bodies. The sequenced genome revealed that this primate belongs to the group of primates that includes monkeys and humans, not the one that include lemurs. Interestingly, a slew of strange insertions—including an entire mitochondrial genome—were discovered in the animal's DNA. The hope is to use this knowledge in efforts to protect species extinction.

THE HUMAN GENOME PROJECT

The **Human Genome Project** (**HGP**) is an international collaborative project which was launched in 1990, setting the ambitious and challenging goal of determining the sequence of the human genome, and ultimately identifying and mapping all human genes. For nearly a decade the proposal was strongly debated, because of the magnitude of effort and expense. Appropriately, the HGP selected the famous drawing "The Vitruvian Man" by Leonardo da Vinci (Fig. 15.2) for its logo. It remains the world's largest collaborative biological project. An initial rough draft of the genome was available in June 2000. Early in 2001 a working draft was published, followed by the final sequencing mapping of the human genome on April 14, 2003. Later analysis of the completed

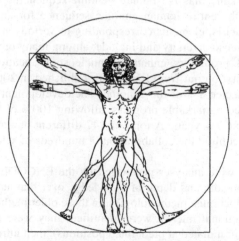

FIG. 15.2 Logo of the Human Genome Project: Vitruvian Man, Leonardo da Vinci. *(From https://commons.wikimedia.org.)*

project indicated that 92% of sampling exceeded 99.99% accuracy, which was within the intended goal.

The HGP took 13 years and cost $2.7 billion. Funding came from the National Institutes of Health (US government agency) and government-sponsored research centers in the United States, the United Kingdom, Japan, France, Germany, Canada, and China. Twenty universities contributed laboratories and human resources. A parallel project, formally launched in 1998, was carried out outside of government by the Celera Corporation.

In May 2016, scientists considered extending the HGP to include creating a synthetic human genome. In June 2016, scientists formally announced **HGP-Write**, a plan to synthesize the human genome. The authors of the proposal asserted that fabricating huge stretches of DNA would allow for numerous scientific and medical advances. The possibilities are endless; for example, it might be possible to make organisms resistant to all viruses, or make pig organs suitable for transplanting into humans. The project will be run by a new nonprofit organization called the Center of Excellence for Engineering Biology, which will raise funds from various public and private sources. As is usually the case with such projects, there are numerous ethical, legal, and social issues that need to be addressed.

ENCODE RESULTS RAISE QUESTION. WHENCE BIOLOGY?

A revolution in Molecular Biology was initiated in 2012 when the full results of the **ENCODE (Encyclopedia of DNA Elements)** project were released. This project had originated shortly after the completion of the HGP. It is essentially an effort to extract as much biological and biochemical information as possible from the massive human genome-sequencing effort. ENCODE aimed to mine the entire human genome sequence for an encyclopedia of **functional elements,** elements corresponding to various genomic functions. These have immense diversity and include, among many others, transcription start sites (TSS), promoters, enhancers, nucleosome locations, and methylation sites. The first output from ENCODE, published in 2007, was an analysis of only a selected 1% of the genome. The development of powerful new methods provided remarkable progress, allowing 100% of the genome to be analyzed the next few years. A total of 147 different human cell types were analyzed. This required the collaboration of hundreds of scientists around the world.

The genome-wide analyses performed by the ENCODE project provided unprecedented breadth and depth of knowledge over that achieved by the use of traditional reductionist methodologies. In these older methods, when genetic elements of functional interest were identified, they were systematically altered (truncated or mutated at predefined positions), then introduced into living cells where their function was investigated by traditional methods. As we have seen, these approaches have yielded a wealth of information on the functions of specific genomic regions of limited lengths and the factors that interact with them, but they must be applied in a case-by-case, time-consuming manner.

Researchers gradually recognized that functional elements had common biochemical or biophysical features (signatures). This led to the development of genome-wide methods, both experimental and computational, that look for these specific biochemical signatures in order to scan whole genomes for the occurrence and distribution of individual functional elements. The biochemical signature strategy became the cornerstone of the ENCODE project's efforts to identify all functional elements in the human genome that are specified by the genomic sequence. The magnitude of this enterprise must be appreciated; it involved repeated scanning over three billion base pairs to search for particular patterns, and then look for correlations between them.

SO, WHAT WAS LEARNED FROM ENCODE?

An enormous amount, not all of which has been digested. Here, we present only a few conclusions relevant to transcription. Some of the findings were expected from the knowledge accumulated from studies of gene-specific systems, whereas others revealed a complex picture of transcription regulation and some complete surprises.

The ENCODE project has systematically mapped features in the genome that relate to transcription: transcribed regions, TSS (Transcription Start Sites), transcription factor (TF) binding sites, chromatin structure and histone modifications, and DNA methylation. The analyses of promoters and TSSs provide information on many related features of the chromatin organization in these elements (nucleosome positioning, histone modification patterns, and so on), while at the same time investigating the connectivity of these elements with other recognized regulatory elements. In addition, the project addressed issues concerned with both the evolutionary conservation of regulatory elements and sequence variations in these elements. The results are of unprecedented magnitude, and impossible to cover even partially here. A few major points:

- Although about 85% of the genome appears to be transcribed into RNA products, only about 3% corresponds to protein-coding mRNAs. Thus, the idea that has permeated molecular biology for so long that DNA functions primarily to code for proteins is WRONG!
- Even taking into account such substances as rRNA and tRNAs, there remains a large fraction of active transcription whose function we are only beginning to guess—the "dark matter" of molecular biology. That much of it may have to do with development is suggested because its abundance is much greater in more complex organisms.
- Chromatin structure is closely related to gene function. Genes that are transcribed almost invariably show a nucleosome-free region right at the transcription start site.

- There is also good evidence for greater DNA accessibility at promoters and TF binding sites.
- The project discovered numerous candidate regulatory elements that are physically associated with one another and with potential for regulating gene expression.
- The ENCODE project revealed a complex network existing between distal transcription control elements and promoters. Most promoters were connected to more than one distal control element, and vice versa, most distal control elements interacted with more than one promoter.
- The ENCODE analysis of the entire human genome confirmed the role TFs play in regulation of eukaryotic transcription. It also revealed novel features of TFs and their cooperation in defining the transcriptional status of a gene, and more broadly in shaping cellular identity.

TFS INTERACT IN A HUGE NETWORK

One of the major findings of the ENCODE project concerns the elements responsible for cell-selective transcription regulation and the complex combinatorial patterns of TFs that are bound to these elements. As stated by John Stamatoyannopoulos "Although ENCODE was conceived as a genomic annotation project fundamentally focused on the linear organization of sequence elements, it is now becoming clear that connectivity between linear elements is an intrinsic part of this annotation—from splicing, to long-range chromatin interactions, to TF networks."

The main conclusion from these studies is that human TFs "co-associate in a combinatorial and context-specific fashion; different combinations of factors bind near different targets, and the binding of one factor often affects the preferred binding partners of others. Moreover, TFs often show different co-association patterns in gene-proximal and -distal regions" (Gerstein et al., 2012).

WHERE IS ENCODE LEADING?

Aside from the many specific revelations mentioned earlier, what are the main lessons obtained to date from ENCODE? It seems to us that they are twofold: first, it is now clear that human cells utilize, in some fashion, much more of the genomic information encoded in their DNA than was hitherto expected. Perhaps it will turn out to be 100% when all cell types have been interrogated. At any rate, "junk DNA" is no longer a useful term. To be sure, we do not know the functional importance of even a fraction of the transcripts that have been identified and, unless new screening methods are devised, will not know for a very long time. But if there is any practical lesson from the history of molecular biology, it is that powerful new methods often appear quickly when needed.

A second surprise is the finding that the vast majority of the newly detected functional entities and interactions among them are cell-type specific. Perhaps this revelation should not have been surprising, given the remarkable number of very different varieties of cells present in the adult human. A complicating aspect of this is that the function of a given gene or regulatory element may vary, depending on the cellular milieu in which it exists. This is going to make the unscrambling of interconnected regulatory pathways even more complex, but additionally rewarding. Perhaps the major impact of this line of research will be in the areas of developmental biology and evolution. There are already hints that the fundamental reason for the differences in size between mammalian and invertebrate genomes lies in this difference in cell-type diversity. Now it can be analyzed and quantified. It may be that a new age in biology has been born.

ATTEMPTS AT A CONTEMPORARY DEFINITION OF A GENE

We have known the existence of genes from the time of Mendel. But our concept as to what they are has steadily changed, and the most recent results from molecular biology continue to modify the concept (for a historic account of the definition of genes, see Table 15.1). This section describes that special history. A more detailed account can be found in Box 15.1.

The golden age of classical genetics culminated with the outstanding studies of fruit flies by Thomas Hunt Morgan. From the work of Morgan and other geneticists, the formal rules for the behavior of genes were established, and by 1930 many gene locations were mapped on chromosomes. Yet the nature of the gene itself was then completely unknown and did not even seem of great importance to the geneticists of the time. Indeed, Morgan stated in his Nobel address (1933), "it does not make the slightest difference whether the gene is a hypothetical unit or whether the gene is a material particle."

That point of view began to change in the 1940s, as rapid advances in biochemistry unveiled the role of proteins, particularly enzymes. It became clear that certain hereditary malfunctions in metabolism, such as albinism, mapped to specific gene locations and were also linked to the malfunction (or lack) of particular enzymes. This led to the idea that genes somehow dictated protein structure—"one gene, one enzyme"—and thus were not only heritable but also participated in the everyday workings of each cell (Table 15.1). The concept was broadened to include proteins other than enzymes with the recognition of sickle-cell anemia as a genetic disease (Fig. 15.3).

A critical breakthrough came with Sanger's sequencing of insulin, in the period 1949–55. The results of Sanger's work showed that the amino-acid sequence of each protein was defined and was unique for that protein. This demonstrated that the genetic material, whatever it was, was able to contain and transmit sequence information to direct the synthesis of proteins. Thus, the genetic substance must be a template of specific sequence, which can not only be inherited (replicated in cell division) but also carries a code of some kind that

TABLE 15.1 Historic Account of the Definition of Genes

Year and Person	Background	Definition
1909 Wilhelm Johannsen	Based on Mendel's concepts	Introduces the term Gene: "Special conditions, foundations and determiners which are present [in the gamete] in unique, separate and thereby independent ways [by which] many characteristics of the organism are specified"
1910 Thomas Morgan	Studied segregation of mutations in *Drosophila*; suggested that genes are arranged linearly and the probability of cross-over is determined by the distance between them	Genes are distinct abstract entities that determine the phenotype; genes can recombine to give rise to new phenotypes
1928 Frederick Griffith	A principle from virulent but dead *Pneumococcus* strain can transform live nonvirulent strains into virulent bacteria	Genes are physical molecules
1941 Beadle and Tatum	Mutations in genes could cause defects in metabolic pathways that are supported by enzymes	"One gene, one enzyme", later was generalized as "one gene, one protein"
1944 Avery, McLeod and McCarty	Investigated the biochemical nature of the transforming principle of Griffith	The principle that can enter into and transform bacterial strains is DNA
1952 Hershey and Chase	Investigated the functions of viral proteins and DNA in the growth of bacteriophage	The factor that enters the bacterial cell during viral infection is DNA, not protein
1953 Watson and Crick	Modeled the structure of DNA as a double helix	Explained how the genetic information encoded in the sequence of bases in the DNA double helix can be copied (replicated) into a daughter DNA molecule or transcribed into an RNA molecule

1958 Crick	Formulated the "Central Dogma" of molecular biology	The central dogma defines the direction of information flow from DNA through RNA to protein; in more general terms, a code residing on nucleic acids gives rise to a functions product (protein or RNA)
1965 Nirenberg and Khorana	Used cell-free systems for synthesis of peptides on chemically synthesized RNA molecules of predefined sequences as templates	Deciphered the genetic code: each amino acid in a polypeptide chain is coded for by a triplet of nucleotides in DNA; the code is universal and nonoverlapping
1970–1990 Numerous labs	Massive cloning and sequencing efforts; development of computational tools for sequence analysis	Genes can be identified based on their sequence characteristics; ORF (open reading frames) are defined as DNA sequences that have the potential to give rise to proteins; this led to the concept of the "nominal" gene which is solely defined by its predicted sequence
1990–2000 Large international organizations/consortia	Analysis of sequenced whole genomes, with emphasis on traditional, protein-coding genes	The Human Genome Nomenclature Organization defines the gene as "a DNA segment that contributes to phenotype/function; in the absence of demonstrated function a gene may be characterized by sequence, transcription or homology" The Sequence Ontology Consortium definition is: "a locatable region of genomic sequence, corresponding to a unit of inheritance, which is associated with regulatory regions, transcribed regions and/or other functional sequence regions"
2003–present ENCODE project	Analysis of the entire human genome for various functional elements	Proposal that "the transcript be considered as the basic atomic unit of inheritance. The term gene would then denote a higher-order concept intended to capture all those transcripts (eventually divorced from their genomic locations) that contribute to a given phenotypic trait."

BOX 15.1   The Classic Definition of a Gene Is Not Consistent With Recent Observations

For a deeper understanding of the problems with the previous definitions of the gene, we list some of the numerous violations to these definitions.

- *Imprinting* (allele-specific expression defined by parent-specific differential DNA methylation) (phenotype is not strictly determined by genotype). Inherited information may not be based in the sequence of nucleic acids.
- *Chromatin structure* (phenotype is not strictly determined by genotype). Chromatin structure influences gene expression: gene expression is controlled by elements/structures beyond nucleotide sequence.
- *Spliced genes* (violates the one-gene, one-protein rule). Alternative splicing between exons of one genetic locus could give rise to multiple different mRNAs, hence, to multiple protein "isoforms" that may have different functions. There were attempts to include this characteristic feature of eukaryotic genes in some definitions. For example, Celera (the company that sequenced the human genome, in parallel with the Human Genome Organization) defines the gene as a "locus of cotranscribed exons"; Ensembl (another international organization that compiles genetic information of vertebrates, including humans) defines the gene as a "set of connected transcripts," sharing at least one exon (but no one exon is common to all of them).
- *Trans-splicing* (ligation of two separate mRNA molecules) (violates the concept of the gene as a "locus"). There are examples of transcripts from the same gene or genes residing on separate chromosomes that are joined to each other before splicing.
- *Tandem chimerism* (aka transcription-induced chimerism; intergenic splicing) (violates the concept of a gene as a "locus"). Two consecutive genes are transcribed into a single long transcript, which is processed to a single mRNA (the intergenic region is being handled as a "normal" intron).
- *RNA editing* (enzymatic modification of bases in RNA) (the structure and the function of the RNA are not exclusively determined by DNA sequence).
- *Protein splicing* (violation): start and end points of proteins are not determined by the gene (i.e., by start and stop codons). Protein product self-cleaves to generate multiple functional products (e.g., viral polyproteins).
- *Protein trans-splicing* (violation): start and end points of proteins are not determined by the gene (i.e., by start and stop codons). Distinct proteins can be spliced together in the absence of a trans-spliced transcript.
- *Protein posttranslational modifications* (methylation, phosphorylation, etc.). The structure and the function of the protein are not exclusively determined by DNA sequence.

specifies protein sequences. With the Watson–Crick model of DNA structure of 1953, together with the experiments of Hershey and Chase (Chapter 7), it finally became clear that the genetic substance must be DNA. At this moment in history, the gene became not just an abstract location on a chromosome, but a part of a macromolecule that codes for a protein.

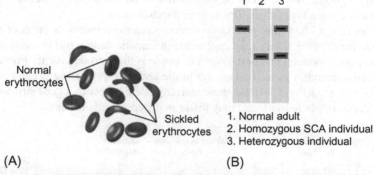

Normal erythrocytes

Sickled erythrocytes

1 2 3

1. Normal adult
2. Homozygous SCA individual
3. Heterozygous individual

(A)   (B)

FIG. 15.3   Sickle cell anemia, a molecular disease. (A) Shapes of normal and sickled erythrocytes. (B) Electrophoretic behavior of hemoglobin from normal individuals, and individuals homozygous and heterozygous for the mutation in the beta chain of hemoglobin.

A current textbook definition of a gene in molecular terms is "The entire nucleic acid sequence that is necessary for the synthesis of a functional polypeptide (or RNA)" (Lodish et al., 2007). We shall see that there are numerous problematic issues with this. These concern gene locations, gene structure, structural variations, and last, but not least, gene regulation.

First, consider overlapping genes. We now know that (1) genes may share the same DNA sequence in a different reading frame or on the opposite DNA strand; (2) one gene may reside entirely within introns of another gene; (3) one gene may overlap with another one on the same strand without sharing any exons or regulatory elements. The existence of overlapping genes violates the one-gene, one-protein rule.

Second, there are large amounts of so-called junk DNA in the genome. Only ~1% of the human DNA codes for exons, that is for proteins. On the other hand, transcription of the genome is pervasive, giving rise to a huge number of noncoding RNAs (large and small). A large amount of these are conserved, suggesting important functional roles. Are these genes? This unexpectedly low ratio of protein-coding regions to the rest of the gnome sequences violates the idea of defined functionality of a gene product, and the connection between genotype and phenotype.

Third, the existence of transposons ("jumping" genes, mobile DNA sequences) violates the idea that the gene is necessarily fixed in one location, "locus." Transposons can be transcribed into RNA which is then reverse-transcribed into DNA to be inserted into the genome DNA as part of their propagation through the genome (this mechanism of propagating transposons is known as retrotransposition, since it involves reverse transcription of information, from RNA into DNA). Many "dead" transposons are transcriptionally "alive."

And then there are pseudogenes: genomic sequences that arise from functional genes but that cannot encode the same type of functional product (i.e., protein, tRNA, or rRNA) as the original genes. Under this definition, retrotransposons would NOT be considered to be pseudogenes because they encode the same type of product as their parental genes.

Fourth, we know that genes can undergo rearrangements as parts of their normal functioning. DNA rearrangements in somatic cells result in many alternative gene products. The best-known example is the rearrangement of portions of immunoglobulin genes that occurs in the antibody-producing B cells of the immune system (Fig. 15.4). So, gene rearrangements show that gene structure is not necessarily hereditary and may differ in different cells/tissues.

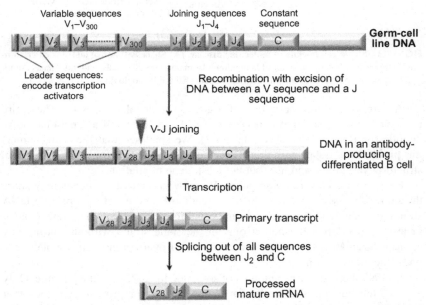

FIG. 15.4   Gene rearrangements for the production of one class of light chains (κ) that constitute a portion of an immunoglobulin molecule. Top, gene organization in a germ-cell line. Germ cells do not produce antibodies: each light chain is encoded by noncontiguous sequences on the same chromosome: variable (V), joining (J), and constant (C). In humans, there are ~300 V sequences, each encoding the first 95 amino acids of the variable region. Each V sequence is preceded by its own leader sequence *(vertical black lines)* that contains transcription activator sequences (not expressed in the germ line); the V sequences form a tight cluster on the chromosome. Each of the four J sequences encodes the last 12 amino-acid residues of the variable region; they from a separate cluster. Finally, there is one C region. During differentiation of one antibody-producing clone of B-cells, gene rearrangement occurs: the final mature mRNA (and the polypeptide) will contain one V, one J, and one C region. The DNA sequences that are excised in the recombination process (in the particular example shown the sequences between V_{28} and J_2) are permanently lost from all progeny of this particular cell line. Note, however, that the V sequences upstream of the junction (in this case V_1–V_{28}) and the downstream J sequences (in this case J_2–J_4) remain in the DNA. The other steps that lead to the production of a functional antibody molecule occur at the level of transcription and further modifications of the mRNA transcript. Transcription uses the leader sequence of only the V sequence that had been joined to the J sequence, thus producing a mRNA precursor that contains only one V sequence. The removal of the extra J sequences occurs during primary transcript splicing. *(From Molecular Biology: Structure and Dynamics of Genomes and Proteomes by Jordanka Zlatanova and Kensal E. van Holde. Reproduced by permission of Taylor and Francis Group, LLC, a division of Informa plc. Copyright 2015.)*

Fifth, there may be differences among individuals in the number of copies of genes or gene regulatory elements. The gene copies may evolve independently, thus affecting evolution.

Finally, the newly established facts about gene regulation violate the one-to-one correspondence between a gene and its regulatory elements. (1) Many regions can regulate one and the same gene in terms of its transcription, mRNA degradation, and posttranslational modifications. (2) One and the same region can regulate several genes. The problem that arises is how to include the gene regulatory regions in a gene definition, especially when they are physically separated by long stretches of DNA?

Based on the existence of these numerous violations to the old definition of a gene (Box 15.1), two major definitions have been recently formulated. (1) "The gene is a union of genomic sequences encoding a coherent set of potentially overlapping functional products" (Gerstein et al., 2007). The definition is based on the discoveries from the ENCODE project; regulatory regions are not included because of many-to-many complex relationships. (2) "The gene is a union of genomic sequences and the regulatory factors associated structurally or functionally with them." (Portin, 2009). This definition is considered "systemic," since it attempts at unifying the concepts of genotype, norm of reaction, and phenotype.

In being truthful to history, we note that that as early as 1986, Rafael Falk, an Israeli geneticist, historian and philosopher of biology stated: "The gene is neither discrete nor continuous, nor does it have a constant location, nor a clear-cut function, not even constant sequence, nor definite borderlines." What an insight! This statement, although correct from the point of view of our contemporary knowledge, cannot be considered strictly a definition, because of all of its assertions are negative.

The extreme challenges in defining a gene will undoubtedly lead to some new definitions in the future. So, the problem is still wide open, as is so much in science.

EPILOGUE

This chapter brings us nearly to the present day in telling what molecular biological techniques are revealing about the fundamentals of genetics and biology. Of course, it is not possible to be truly contemporary in such a dynamic field. It is clear that we are at the verge of a whole new understanding of the regulation of gene expression. The problem of the "dark" genes that are expressed but do not code for proteins is a major challenge. It seems likely that this will bring major new insights into the mechanisms of development and cell differentiation, which in turn will aid in our understanding of disease state, including cancer. There is still so very much to learn!

FURTHER READING

Books and Reviews

Chanock, S., 2012. Toward mapping the biology of the genome. Genome Res. 22, 1612–1615. An overview of the series of papers published in the same issue of Genome Research (see example papers below). The success of these papers should be measured by the scope of the scientific insights and tools, but also by their ability to attract new talent to mine the complex data sets produced.

Ecker, J.R., Bickmore, W.A., Barroso, I., Pritchard, J.K., Gilad, Y., Segal, E., 2012. Genomics: ENCODE explained. Nature 489, 52–55. Six scientists describe in individual forum articles the project and discuss how the data are influencing research directions across many fields.

Frazer, K.A., 2012. Decoding the human genome. Genome Res. 22, 1599–1601. This commentary describes some of the highlights of the ENCODE project, including technical accomplishments, high quality data sets, and integrated analyses with other resources, such as disease-associated variants identified through genome-wide association studies.

Gerstein, M.B., Bruce, C., Rozowsky, J.S., Zheng, D., Du, J., Korbel, J.O., Emanuelsson, O., Zhang, Z.D., Weissman, S., Snyder, M., 2007. What is a gene, post-ENCODE? History and updated definition. Genome Res. 17, 669–681. An in-depth review of the evolution of operational definitions of a gene over the past century—from the abstract elements of heredity of Mendel and Morgan to the present-day open-reading frames enumerated in the sequence databanks. A tentative update to the definition of a gene is proposed.

Gerstein, M.B., Kundaje, A., Hariharan, M., Landt, S.G., Yan, K.K., Cheng, C., Mu, X.J., Khurana, E., Rozowsky, J., Alexander, R., Min, R., Alves, P., Abyzov, A., Addleman, N., Bhardwaj, N., Boyle, A.P., Cayting, P., Charos, A., Chen, D.Z., Cheng, Y., Clarke, D., Eastman, C., Euskirchen, G., Frietze, S., Fu, Y., Gertz, J., Grubert, F., Harmanci, A., Jain, P., Kasowski, M., Lacroute, P., Leng, J.J., Lian, J., Monahan, H., O'Geen, H., Ouyang, Z., Partridge, E.C., Patacsil, D., Pauli, F., Raha, D., Ramirez, L., Reddy, T.E., Reed, B., Shi, M., Slifer, T., Wang, J., Wu, L., Yang, X., Yip, K.Y., Zilberman-Schapira, G., Batzoglou, S., Sidow, A., Farnham, P.J., Myers, R.M., Weissman, S.M., Snyder, M., 2012. Architecture of the human regulatory network derived from ENCODE data. Nature 489, 91–100. Genomic binding information of 119 transcription-related factors in over 450 distinct experiments have been analyzed. The combinatorial, co-association of transcription factors was found to be highly context specific: distinct combinations of factors bind at specific genomic locations.

Lander, E.S., et al., 2001. Initial sequencing and analysis of the human genome. Nature 409, 860–921. The paper reports results of an international collaboration to produce and make freely available a draft sequence of the human genome. Initial analysis of the data is also provided.

Lodish, H., Berk, A., Kaiser, C.A., Krieger, M., Scott, M.P., Bretscher, A., Ploegh, H., Matsudaira, P., 2007. Molecular Cell Biology, sixth ed. Freeman and Co, New York. Contemporary authoritative textbook covering landmark experiments, with emphasis on medical relevance.

Portin, P., 2009. The elusive concept of the gene. Hereditas 146, 112–117. A brief review of recent observations and discussion of the difficulties of defining the gene. Stresses the fact that the semantic information content of genes is context-dependent: genes assume their biochemical characteristics only within living cells, their developmental characteristics only within living organisms, and their evolutionary characteristics only within populations of living organisms.

Stamatoyannopoulos, J.A., 2012. What does our genome encode? Genome Res. 22, 1602–1611. A discussion of many of the leading findings of the ENCODE project with vision of the future in terms of maximizing the accuracy, completeness and utility of ENCODE data for the community.

Chapter 16

Whole Genomes and Evolution

PROLOGUE

Whenever a radically new scientific methodology arises, it stimulates advances in related sciences and technology. That has been surpassingly true in the case of molecular biology. In these last two chapters, we will describe some ways in which other fields of biology, medicine, and technical applications have stemmed from the insights and methods we have surveyed throughout the book. As an example of scientific cross-fertilization, we choose evolution—a science long on the perimeters of biology. To show its connection to molecular biology, we first briefly review the history of this field.

EVOLUTIONARY THEORY: FROM DARWIN TO THE PRESENT DAY

Vague ideas that more complex creatures somehow evolved from simpler forms persisted from Aristotle in ancient Greece to Jean-Baptiste Lamarck in the early 19th century. Lamarck, in the first dynamic model for evolution, proposed that creatures took the forms that their environment required them to assume. Although the mechanism for such adaptation was obscure (and incorrect), it remained the predominant view for much of the 19th century. Then, the whole field was overturned by Charles Darwin in 1859.

Darwin was born in 1809 in a wealthy educated family and was interested in natural history from his early childhood. He accepted an invitation for a self-funded position of naturalist on the ship H.M.S. Beagle, whose main objective was to chart the coastline of South America. The trip lasted for almost 5 years, 1831–36, and provided ample time and resources for the young naturalist to study local geology and collect samples, mainly beetles and marine invertebrates. On the way home, he organized his extensive notes, writing that his observations "would undermine the stability of Species."

The publication, in 1859, of his book "On the Origin of Species" marked the beginning of a new era of understanding of life and evolution. It also caused a furor, for it denied the "separate creation" of each species, which had been a major tenet of Christian theology. The anger was amplified by the publication a few years later of "The Descent of Man," which specifically proposed the evolution of humans from great apes. Fig. 16.1 lists some of the most important

The Evolution of Molecular Biology. https://doi.org/10.1016/B978-0-12-812917-3.00016-4

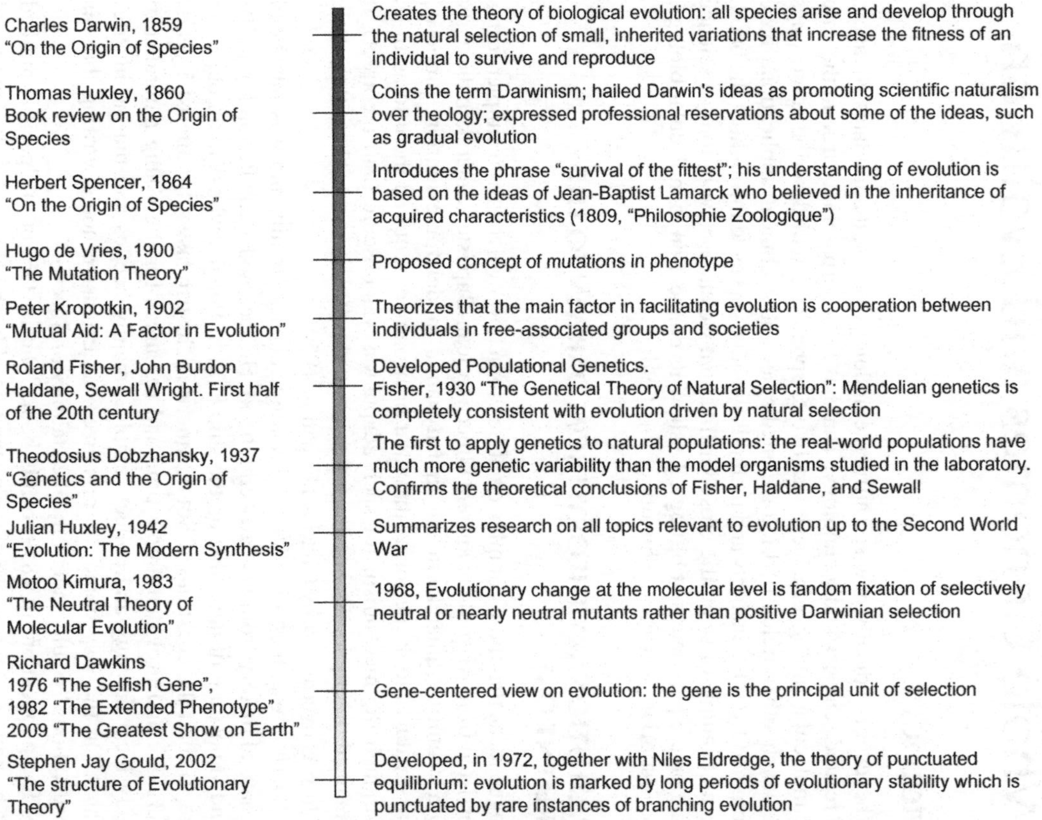

Creates the theory of biological evolution: all species arise and develop through the natural selection of small, inherited variations that increase the fitness of an individual to survive and reproduce

Coins the term Darwinism; hailed Darwin's ideas as promoting scientific naturalism over theology; expressed professional reservations about some of the ideas, such as gradual evolution

Introduces the phrase "survival of the fittest"; his understanding of evolution is based on the ideas of Jean-Baptist Lamarck who believed in the inheritance of acquired characteristics (1809, "Philosophie Zoologique")

Proposed concept of mutations in phenotype

Theorizes that the main factor in facilitating evolution is cooperation between individuals in free-associated groups and societies

Developed Populational Genetics.
Fisher, 1930 "The Genetical Theory of Natural Selection": Mendelian genetics is completely consistent with evolution driven by natural selection

The first to apply genetics to natural populations: the real-world populations have much more genetic variability than the model organisms studied in the laboratory. Confirms the theoretical conclusions of Fisher, Haldane, and Sewall

Summarizes research on all topics relevant to evolution up to the Second World War

1968, Evolutionary change at the molecular level is random fixation of selectively neutral or nearly neutral mutants rather than positive Darwinian selection

Gene-centered view on evolution: the gene is the principal unit of selection

Developed, in 1972, together with Niles Eldredge, the theory of punctuated equilibrium: evolution is marked by long periods of evolutionary stability which is punctuated by rare instances of branching evolution

Charles Darwin, 1859
"On the Origin of Species"

Thomas Huxley, 1860
Book review on the Origin of Species

Herbert Spencer, 1864
"On the Origin of Species"

Hugo de Vries, 1900
"The Mutation Theory"

Peter Kropotkin, 1902
"Mutual Aid: A Factor in Evolution"

Roland Fisher, John Burdon Haldane, Sewall Wright. First half of the 20th century

Theodosius Dobzhansky, 1937
"Genetics and the Origin of Species"

Julian Huxley, 1942
"Evolution: The Modern Synthesis"

Motoo Kimura, 1983
"The Neutral Theory of Molecular Evolution"

Richard Dawkins
1976 "The Selfish Gene",
1982 "The Extended Phenotype"
2009 "The Greatest Show on Earth"

Stephen Jay Gould, 2002
"The structure of Evolutionary Theory"

FIG. 16.1 Timeline of some of the most important developments in the field of biological evolution.

developments in the field of biological evolution, reflecting the struggle of ideas and the changes in paradigms that stemmed from new observations in nature and in the laboratory.

It must be understood that Darwin's great contribution was not the *idea* of evolution. That was old. Rather, he proposed a *mechanism*—**natural selection** of the fittest.

Four preconditions are needed for the process Darwin called natural selection. If these conditions are met for any property of a species, natural selection automatically results. We will borrow well-written descriptions from Mark Ridley's book "Evolution" (2004).

1. Reproduction. Entities must reproduce to form a new generation.
2. Heredity. The offspring must tend to resemble their parents: roughly speaking, "like must produce like."
3. Variation in individual characters among the members of the population. If we are studying natural selection for body size, then the population must have a range of body sizes.
4. Variation in the *fitness* of organisms according to the state they have for a heritable character. In evolutionary theory, fitness is a technical term, meaning the average number of offspring left by an individual relative to the number of offspring left by an average member of the population. This condition therefore means that individuals in the population with some characters must be more likely to reproduce (i.e., have higher fitness) than others.

Natural selection can explain both evolutionary change and the absence of change. If the environment is constant and no superior form arises in the population, natural selection will keep the population unchanged. Natural selection also explains adaptation.

Natural selection can be categorized in three forms: stabilizing, directional, and disruptive. We illustrate these types on a trait that is continuously distributed, such as body size (Fig. 16.2). In all three forms of selection, the frequency distribution of body size in the population is the normal (bell-shaped) distribution, and the same (middle panels). The most common case for natural selection is **stabilizing**. The average members of the population, with intermediate body sizes, have higher fitness than the extremes. Natural selection now acts against changes in body size and keeps the population constant through time. Some scientists refer to this type of selection as negative (or purifying) selection because it occurs through selective removal of gene alleles (and thus, traits) that are deleterious.

We can illustrate **directional** selection by assuming that that smaller individuals have higher fitness (i.e., produce more offspring) than larger individuals. This will favor smaller individuals and will, if the character is inherited, produce a decrease in average body size with time. Directional selection could, of course, also produce an evolutionary increase in body size if larger individuals had higher fitness.

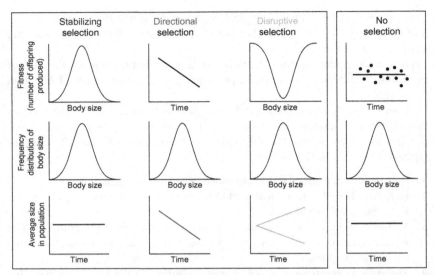

FIG. 16.2   Charts illustrating the different types of genetic evolution on the example of body size. See text for further explanation. *(From OUR ORIGINS: DISCOVERING PHYSICAL ANTHROPOLOGY, SECOND EDITION by Clark Spencer Larsen. Copyright © 2011, 2008 by W.W. Norton & Company, Inc. Used by permission of W. W. Norton & Company, Inc.)*

The third type of natural selection, **disruptive** selection, occurs when both extremes are favored in terms of fitness relative to the intermediate types. Selection will lead to fixing both extremes with time.

It is important to understand that natural populations show variation at all levels, from DNA sequences to gross morphology. New variation is random with respect to the direction of adaptation. Adaptively directed mutation, as was proposed by Lamarck, is unlikely, for theoretical reasons. "It is theoretically difficult to see how any genetic mechanism could have the foresight to direct mutations in this way" (Ridley, 2004). But there was a major problem with Darwin's depending on gradual variation to drive evolution. It would be agonizingly slow! To actually generate new species in this way would take much longer than the age of the earth, estimated in Darwin's time to be between 20 and 100 million years. A way out of the dilemma was suggested by the Dutch biologist Hugo DeVries, who originated, around 1900, the idea of **mutations**, sudden changes in phenotype. We now ascribe phenotypic mutations to changes in DNA, transmitted through gene expression (see Chapter 5). The Central Dogma of molecular biology fits nicely with the theory of evolution, as developed by Darwin and DeVries.

There is one other point about evolution that must be emphasized, for it is so frequently misunderstood. Natural selection is not working toward some goal of perfection. It is not designed, it is not aimed. Mutations and genome rearrangements happen; if they produce a more fit creature, its chances of propagating are enhanced. That is all.

CLASSIFYING ORGANISMS: PHYLOGENETICS

One of the first stages in any new science is the gathering of data and establishing a rational method for classifying them. Although others had made attempts, the first rational method for classifying organisms according to their similarities and differences was introduced by Carl Linnaeus in the 18th century. Linnaeus' systematics aimed at creating a system to place all known organisms into a logical classification, the *Systema Naturae*. His system divided organisms into a hierarchic series of taxonomic categories, starting with kingdom and progressing down through phylum, class, order, family, and genus to species.

The naturalists of the 18th and early 19th centuries likened Linnaeus hierarchy to a "tree of life," an analogy later adopted by Darwin in *The Origin of Species* as a means of describing the interconnected evolutionary histories of living organisms. Thus, Linnaeus's systematics became reinterpreted as a phylogeny indicating not just the similarities between species but also their evolutionary relationships.

The relevant data for constructing taxonomic schemes or infer phylogenetic relations were traditionally obtained by examining variable characters in the organisms being compared. In the days of Linnaeus these characters were morphological features; such classification suffered from the lack of precision inherent in describing morphology. At present the foundation for these types of analyses are quantitative molecular data, mainly protein and DNA sequences.

PHYLOGENETICS GOES MOLECULAR

The most incisive way to study phylogenetics and trace evolution is to follow the changes in the sequences of proteins and nucleic acids that occur as species evolve. Both comparisons of living, related organisms and analysis of the DNA or protein sequences from fossils provide quantitative measurement of evolution. DNA, and to a lesser extent proteins, has been found to survive over geologic time. Use of PCR allows amplification of minute amounts of DNA. Equally important are novel phylogenetic methods that have been developed to analyze large molecular datasets by rigorous mathematical procedures. The sequences of protein and DNA molecules provide the most detailed and unambiguous data for molecular phylogenetics and techniques for rapid and inexpensive protein and nucleic acid sequencing became routine in the period between 1960 and 1980.

Protein sequences are still used today in some, rather specific, contexts, but DNA sequences have now become by far the predominant choice. This is partly because DNA survives fossilization better than protein, but mainly because DNA yields more phylogenetic information than protein. The nucleotide sequences of a pair of homologous genes have higher information content than the amino acid sequences of the corresponding proteins. Recall from Chapter 10 that several different mutations in a nucleotide sequence may result in no change in peptide sequence due to the degeneracy of the genetic code. This is illustrated in Fig. 16.3.

```
GCC GCA TTC AGA GGU ATA GGA
Ala Ala Phe Arg Gly Ile Gly
                         ↓
Ala Ala Phe Arg Gly Leu Gly
GCC GCA TTT AGA GGU TTA GGC
```

FIG. 16.3   **DNA yields more phylogenetic information than protein.** The two nucleotide sequences in the DNA differ at three positions *(bold letters)*, but the encoded amino acid sequences differ at only one position *(bold letter)*. This is because two of the nucleotide substitutions are synonymous, that is, code for the same amino acid, and one is nonsynonymous.

RNA molecules that do not code for proteins (noncoding RNAs) have been another important tool in inferring evolutionary relationships. Beginning in the late 1970s, Carl Woese and coworkers performed a large-scale comparison among the 16S rRNA of numerous species (the RNA molecule that comprises the RNA scaffold of the small ribosomal subunit). Based on such analysis Woese proposed the existence of prokaryotic organisms, originally called *Achaebacteria*, and later *Archaea*, that differ from bacteria. Woese redrew the tree of life, defining three domains, *Bacteria*, *Archaea*, and *Eukarya* (see Chapter 12).

THE COMPARATIVE GENOMICS REVOLUTION

The sequencing of the human genome published by the International Genome-Sequencing Consortium in 2001 has opened immense opportunities to produce large amounts of sequencing data cheaply and effectively. The technical developments that had to be developed to sequence the 3.2 billion nucleotides in the human nuclear DNA were further refined to the extent that we are now in possession of thousands of individual human genomes and complete genomes of an ever-increasing number of different biological species, from viruses and bacteria, through fungi and plants, to invertebrates, to primates. Remarkably, this list includes the extinct hominid, *Homo neanderthalensis*.

Comparative genomics can be applied to numerous questions. One important application is to ask if a sequence found in two genomes is conserved beyond neutral expectations. The conservation of a sequence suggests that it has evolved under selective pressure because of some important role it plays. Such analysis is particularly revealing for sequences outside of those that code for proteins; the latter are expected to be under selective constraints. As put by Alföldi and Lindblad-Toh (2013) here is where comparative genomics "shines." A related question concerns gene loss: if a sequence is expected to be conserved, based on its known role, but is not, this may be interpreted in terms of phenotypic divergence. Another important problem that scientists are looking into is changes in the location of a gene, due to transpositions, translocations, and other forms of chromosomal rearrangements. Changes in gene location may confer adaptive advantages.

TRACING HUMAN EVOLUTION

We will present two curious examples of using comparative genomics to solve interesting problems concerning an issue of interest to all human beings: human evolution. The first one concerns the evolution of the sex chromosomes (Fig. 16.4). Sex-linked genes are arranged in many different ways over species. The two sex chromosomes present in humans, the X and the Y chromosome, originated from a homologous pair of chromosomes. The current male Y chromosome has lost so many genes in comparison to its female X counterpart that some researchers believe that it is destined to disappear altogether. Comparative genomics, however, show a different picture. Sequencing of the chimpanzee Y chromosome shows rapid evolution which comprises not only gene loss but also addition and duplication of novel sequences. Rhesus macaque Y chromosome analysis did not show gene loss from the homologous chromosomes predecessors of X and Y. The new theory of sex chromosome evolution based on these genomic data suggests that the Y chromosome has indeed lost considerable number of genes, but these were all involved in the ability to cross-over with a homologous partner; the original genes that remain and define sex are subject to strict purifying selection, meaning that the Y chromosomes will continue to exist.

Another interesting issue concerning the evolution of modern humans, including development of speech, was addressed by comparing the human genome with those of the Neanderthals, the chimpanzee and the orangutan.

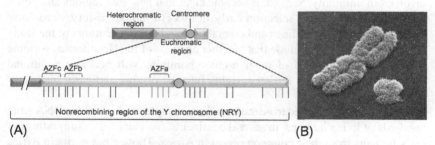

(A)   (B)

FIG. 16.4   The structure of the human Y chromosome. (A) The Y chromosome contains two major regions. The first region can recombine with the X chromosome and exchange portions with it. The second region, called nonrecombining region of the Y chromosome, NRY, cannot exchange regions with the X chromosome. *Light gray bar*, expressed euchromatic portion NRY; gray bar, heterochromatic portion of the NRY; gray oval, centromere. Genes are marked by vertical bars. Green bars indicate genes that are widely expressed housekeeping genes (genes vital for cellular function). Three regions, AZFa, AZFb, and AZFc (azoospermia factor regions a, b, c) are often be deleted in infertile men. Males suffering from this infertility condition, which affects around 1% of the male population, have no measurable level of sperm in semen. (B) Micrograph illustrating the morphology and the size of the two human sex chromosomes. *((A) Based on Lahn, B.T., et al., 2001. Nat. Rev. Genet. 2, 207–216, Fig. 1; (B) From Andrew Syred, with permission from Science Source.)*

The comparison focused on the *FOXP2* human gene, which is known to be mutated in patients with linguistic and grammatical impairments. Two recent nonsynonymous changes in the human *FOXP2* gene are viewed as contributing to the orofacial movement control that privies us with the ability to speak.

The publishing of a draft of the Neanderthal genome in 2010 gave answers to other lingering questions about human evolution. The sequence was derived from analysis of nine DNA extracts from three bones from a cave in Croatia and compared to similar samples from three other sites in Europe and to genomes from five present-day humans. Such studies are technically challenging for several reasons: (1) DNA extracted from bones is invariably degraded to small pieces, average size of less than 200 bp; (2) it contains chemical modifications; (3) the majority of the DNA (95%–99%) comes from microbial organisms that colonized the bones after death; (4) there is significant contamination (11%–40%) with modern human DNA, presumably arising from handling the samples. So, considerable effort was put into enriching the samples for Neanderthal DNA and using clean-room environments. In total, 5.3 Gb of Neanderthal DNA sequence was generated from about 400 mg of bone powder removed from under the bone surface with sterile dentistry drills. Among the numerous inferences derived from this work, two have attracted wide interest. First, it is now beyond doubt that Neanderthals interbred with anatomically modern humans, thus putting an end to a long-standing controversy about the possibility of interbreeding during the periods of coexistence of the two *Homo* species. Among the genes transferred during interbreeding were some of those involved in immunity. Second, it became clear that genomic regions and genes that underwent positive selection early in modern human history were those involved in cognitive abilities and cranial morphology. The authors of the study (Green et al., 2010) conclude that "further analysis of the Neanderthal genome as well as the genomes of other archaic hominins will generate additional hypotheses and provide further insights into the origins and early history of present-day humans."

This has indeed proved to be the case. For example, analysis of the DNA from hundreds of living humans, in several continents, has clarified the migrations of early humans from their common origin in Africa. Those from northern Africa and the Middle East diverged over Europe and much of Asia in one wave, about 70,000–100,000 years ago. In that migration, they interbred with Neanderthals already occupying those areas, accounting for the presence of Neanderthal sequences in Eurasian genomes. The humans who occupied sub-Saharan Africa did not take part in the great trek, and their descendants have no such sequences. Finally, it is now definitely established by such studies that native American peoples descend from the Asian component of this great migration.

Thus, combining data from a variety of disciplines, including, but not limited to, anthropology, archeology, paleontology, neurobiology, evolutionary psychology, embryology, and genetics, evolutionary biologists have arrived at the tree of human evolution presented in Fig. 16.5. A drawing showing the evolution of the skeleton is presented in Fig. 16.6, and a reconstruction of human faces at different stage of evolution is given in Fig. 16.7.

It is beyond doubt that comparative genomics will play a major role in understand evolution. It is also expected to contribute significantly to understanding, treatment and curing of genetic diseases. Could those who first developed cloning in bacteria have guessed how broad the scientific import of their work would be?

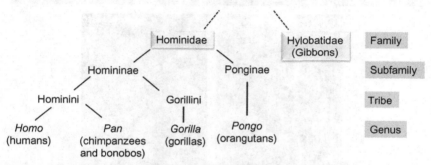

FIG. 16.5   Family tree showing the extant (living) hominoids. All except gibbons are hominids. *(Adapted from Fred the Oyster, Creative Commons, Wikimedia.)*

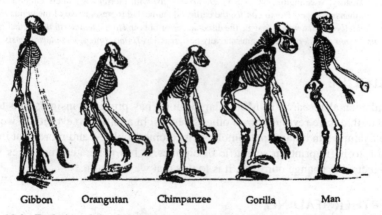

FIG. 16.6   Evolution of the skeleton in apes and humans. The evolution of the skeleton involved the legs, pelvis, vertebral column, feet and ankles, and skull. The knee and ankle joints became increasingly robust and thus can better support increased weight. In the feet, the big toe moved into alignment with the other toes, which is an aid in forward locomotion. In addition, the human S-shaped vertebral column better supports the increased weight on each vertebra in the upright position. The arms and forearms being shortened relative to the legs make it easier to run. *(Adapted from Tim Vickers, commons.wikimedia.org.)*

FIG. 16.7   Reconstructions depicting the early human beings. "Lucy," a female of the genus *Australopithecus* that became extinct 2 million years ago (on exhibit in the National Museum of Natural History, Washington, DC, USA). *Homo erectus*, with morphology intermediate between that of humans and apes. *Homo habilis,* the earliest documented representative of the genus *Homo. Homo heidelbergensis,* which may be the direct ancestor of both *Homo neanderthalensis* and *Homo sapiens. Homo neanderthalensis,* alternatively designated as *Homo sapiens neanderthalensis.*

EPILOGUE

Fundamental research in Molecular Biology has provided insights and techniques that have revolutionized other sciences. In addition to evolution, we can cite examples in botany, paleontology, medicine, zoology, and more. But in addition to its imprint on scientific disciplines, it has generated industries with products that change our lives. It is to this that we now turn.

FURTHER READING

Books and Reviews

Alföldi, J., Lindblad-Toh, K., 2013. Comparative genomes as a tool to understand evolution and disease. Genome Res. 23, 1063–1068. Makes the point that the sequencing of vertebrate genomes helps in the interpretation of the human genome and provide large-scale studies of genome evolution. The comparative genomic data are also of wide use in human medicine.

Dawkings, R., 2009. The Greatest Show on Earth. Free Press/Transworld, New York, NY/London. This book is the author's personal summary "of the evidence that the 'theory' of evolution is actually a fact — as incontrovertible a fact as any in science".

Doolittle, W.F., 2000. Uprooting the tree of life. Sci. Am. 282, 90–95. An attractive early hypothesis of the tree of life (the single tree) suggested the collection of gene sequences and their analysis with the methods of molecular phylogeny. The data show the model to be too simple and new hypotheses are called for.

Futuyma, D.J., 2013. Evolution, third ed. Sinauer Associates, Sunderland, MA. Addresses major themes in contemporary evolutionary biology—including the history of evolution, evolutionary processes, adaptation, and evolution as an explanatory framework—at levels of biological organization ranging from genomes to ecological communities.

Green, R.E., et al., 2010. A draft sequence of the Neandertal genome. Science 328, 710–722. A draft sequence constructed from partial sequencing data of three individuals. Comparisons of the Neandertal genome to the genomes of five present-day humans from different parts of the world identify genomic regions affected by positive selection in ancestral modern humans; these regions include genes involved in metabolism and in cognitive and skeletal development.

Mayr, E., 2002. What Evolution is. Science Masters Series, Basic Books, New York, NY. Gathering insights from his seven-decade career, the renowned biologist Ernst Mayr argues that evolution is now to be considered not a theory but a fact.

National Academy of Sciences, Institute of Medicine, 2008. Science, Evolution and Creationism. . Group of experts assembled by the National Academy of Sciences and the Institute of Medicine explain the fundamental methods of science, document the overwhelming evidence in support of biological evolution, and evaluate the alternative perspectives offered by advocates of creationism, including 'intelligent design'. Free PDF download available.

Noonan, J.P., 2010. Neanderthal genomics and the evolution of modern humans. Genome Res. 20, 547–553. Critical for the utility of a Neanderthal genome sequence is the evolutionary relationship of humans and Neanderthals. Modern human and Neanderthal lineages diverged before the emergence of contemporary humans.

Ridley, M., 2004. Evolution, third ed. Wiley-Blackwell Publishing, Hoboken, NJ. The premier undergraduate text in the study of evolution. In-depth, yet readable and thought-provoking. Provides a historical perspective but also covers the most recent developments in the field.

Chapter 17

Practical Applications of Recombinant DNA Technologies

PROLOGUE

Molecular biology developed as a "pure" science, an attempt to understand biology more deeply. In the preceding chapter, we showed how this new field has had a major impact upon a number of other related sciences. But this is far from the whole story. As is almost always the case, following such fundamental scientific advances, clever people have quickly discovered practical applications of the new concepts. Today, the diversity of methods and insights originating from molecular biological research has generated a vast and ever-growing field, with innumerable effects on our daily lives. It has made it easier to catch criminals (and exonerate the innocent), create new medicines, improve crops, and even edit the genetic makeup of animals. In this chapter, we can only outline a few examples of these applications; it would require many volumes to describe them adequately.

CATCHING CRIMINALS AND FREEING THE INNOCENT

In 1982, the British geneticist Alex Jeffreys was studying small repetitive DNAs, a class of DNA sequences that do not code for protein. In comparing the gel electrophoretic patterns of these DNAs in different humans, he made the remarkable observation that they seemed to be unique for each person. This, he realized, could provide a molecular "fingerprint" much less ambiguous than the traditional fingerprints that had been the mainstay of forensic medicine for decades. The technique developed was termed **DNA profiling**, also known as **DNA fingerprinting** or **DNA typing**.

A few years later, Richard Buckland, a 17-year-old youth with learning disabilities, was convicted of the rape and murder of two young women. The evidence was circumstantial, but under questioning Buckland confessed and was convicted. However, Jeffreys analyzed DNA from semen samples collected from the two crime scenes; they were identical; indeed, one person had committed both crimes—but it was not Buckland! At that time, the technique was absolutely new, so Buckland could not be exonerated on that basis alone. However, a tip from an overheard conversation led to suspicion of one Colin Pitchfork.

The Evolution of Molecular Biology. https://doi.org/10.1016/B978-0-12-812917-3.00017-6

When his DNA was analyzed, an identical match to the crime samples was found! Pitchfork confessed and was sentenced; Buckland was finally exonerated. This was but the first of many such cases. In the United States alone, hundreds of incorrectly convicted individuals have been exonerated by this technique. On the other hand, because complete identity of DNA profiles is difficult to prove, such evidence cannot be used to convict. Apparent identity is, however, useful in directing investigations.

DNA profiling is now the most reliable method for identifications of individuals involved in criminal justice, paternity cases, and immigration disputes. Numerous databases of DNA fingerprints have been complied and are in use worldwide.

The method uses cell samples collected by buccal swabs, blood or semen samples, skin, and hairs. Personal items such as a toothbrush or razor could also be used to obtain samples containing bits of tissue. The DNA is extracted, fragmented into small pieces using restriction enzymes, and subjected to a variety of analytical techniques to look for differences in that 0.1% of the DNA which differs between individuals. The other 99.9% of human DNA sequences are the same in every person. Everyone, however, contains short repetitive sequences in the noncoding regions of the genome; the number of these is highly variable among humans. These are termed variable number tandem repeats (VNTRs). The different number of repeats in individual people gives rise to restriction fragments of different lengths (Fig. 17.1A), which, in the original version of the technique, were identified following gel electrophoresis, hybridization with radioactive probes corresponding to the repeats, and autoradiography (Fig. 17.1B). This method was termed Restriction Fragment Length Polymorphism (RFLP). Today, the analytical method of choice is the Amplified Fragment Length Polymorphism (AFLP), which combines the Polymerase Chain Reaction (PCR, see Chapter 14) with RFLP. Because of the remarkable ability of PCR to multiply minute amounts of DNA, the sample size needed using this technique can be incredibly small. PCR can amplify a DNA region of interest over a million-fold in about 2 h. The whole of this amplification can be carried out in a small instrument called a thermocycler.

The reliability of these methods to identify individuals from small amounts of tissues has been even more a boon to the innocent as it is a threat to the criminal. There are today hundreds of innocent persons walking free who had been erroneously convicted of crimes on the basis of insufficient evidence, later refuted by these techniques. The development of DNA profiling is a wonderful example of basic research leading to far-reaching practical applications.

Very recently, a new DNA-based forensic technique, called **DNA phenotyping**, has received attention. Performing high-speed sequencing of the genomic DNA from the crime scene, this method uses variations in DNA sequence that determine phenotypic markers like eye color or facial shape to construct a predicted image. The method is much more labor-intensive than DNA profiling and has yet to be proved wholly reliable.

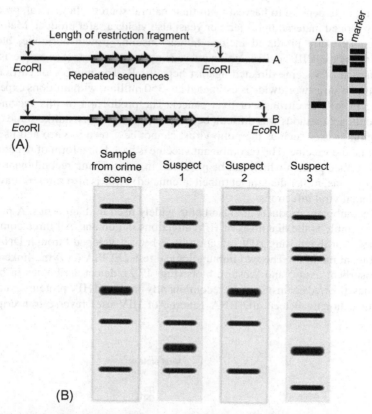

FIG. 17.1   DNA fingerprinting via analysis of VNTRs, variable number tandem repeats. In the noncoding regions of the genome, sequences of DNA are frequently repeated, giving rise to so-called VNTR. The number of repeats varies between different people and can be used in DNA fingerprinting. In the example shown in (A), person A has only four repeats while person B has seven. When their DNA is cut with the restriction enzyme *Eco*RI, which cuts the DNA at either end of the repeated sequence, the DNA fragment produced by B is longer (see electrophoretic behavior of these two fragments on the right). The lane marked M contains marker pieces of DNA that help us to determine the sizes of the DNA fragments in the samples. If lots of pieces of DNA are analyzed in this way, a "fingerprint" comprising DNA fragments of different sizes, unique to every individual, emerges. (B) An example of tentative identification of a suspect (#2) whose DNA profile matches the one recovered from the crime scene. DNA from other suspects does not match. *(Adapted from http://geneed.nlm.nih.gov; courtesy of the National Human Genome Research Institute.)*

PRODUCTION OF PHARMACEUTICAL COMPOUNDS IN RECOMBINANT BACTERIA OR YEAST

This is today the most advanced field of application of recombinant DNA technology. The list of compounds used in clinical practice already encompasses hundreds of products, most of which are proteins. In many cases, these were difficult or expensive to harvest from their natural sources. In general, production of cloned material in bacteria or yeast also yields a safer product. Materials so commercially produced include human insulin, growth hormone, blood-clotting factor VIII, and recombinant vaccines. An important example is the production of vaccine directed against hepatitis B. The number of carriers of hepatitis B virus worldwide is estimated at ~350 million, with millions expected to die from liver cirrhosis or liver cancer. The production of this vaccine by conventional methods was difficult because hepatitis B virus, unlike some other common viruses such as the polio virus, cannot be grown in vitro and used to generate the vaccine. The recombinant vaccine is based on a form of hepatitis B virus surface protein which can be produced in yeast. Using recombinant viral protein alone, rather than intact (albeit attenuated) virus is also safer as it avoids accidental viral infections.

Recombinant products are also being widely used in diagnostics. A prominent example is the diagnosis of HIV infections in humans. All three common methods for diagnosing HIV infection have been developed through DNA recombinant methods. The two immunological tests, ELISA (enzyme-linked immunosorbent assay) and Western blots (Fig. 17.2), detect antibodies in blood samples directed against specific recombinantly produced HIV proteins. Another test that directly detects the RNA genome of HIV uses reverse transcriptase

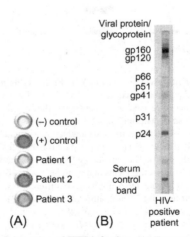

(A) (B)

FIG. 17.2 Immunological detection of HIV infection. (A) Enzyme-linked immunosorbent assay (ELISA) test. Blood samples from three patients were analyzed by ELISA. The objects shown are sample wells in which the reaction is performed, one for each patient, and two for control reactions. In case of positive ELISA (patients 2 and 3), a Western Blot test is performed. (B) Western blot test. A mixture of recombinant HIV proteins is separated by gel electrophoresis. Blood samples from patients are incubated with the separated proteins; if a patient is positive for HIV, the blood will contain antibodies to at least some of the HIV proteins on the gel. The consensus among clinicians is that to be conclusive, the profile must have at least five bands. *((B) Adapted from Suthon, V. et al., 2002, BMC Infect. Dis. 2, 19, Fig. 1, with permission from BioMed Central.)*

polymerase chain reaction (RT-PCR, see Chapter 14). The development of the RT-PCR test was made possible by the molecular cloning and sequence analysis of the HIV genome.

GENETIC ENGINEERING OF PLANTS

Plant genetic engineering is still a field in flux (and not without controversy) despite its many achievements to date. The potential is huge since genetically engineered (GE) plants can provide the world population with foods of high nutritional value. Crop yields can be increased and agricultural crops that are resistant to herbicides and insecticides can be introduced, allowing easy control of insect pests and weeds. These goals are usually achieved by the introduction of one or more genes, from a variety of sources, into the plant genome. Many economically important crops have been engineered to show resistance to the commonly used herbicide glyphosate (commercial name Roundup). Other crops are made resistant to insects by introducing a recombinant form of a bacterial gene whose protein product is selectively toxic to insects. For example, a gene from the bacterium *Bacillus thuringiensis* encodes the so-called Bt toxin, which kills insects. The use of recombinant plants carrying the *Bacillus* toxin gene in their genomes replaces the traditional way of fighting insect predators by spraying fields with the bacterium using aircraft.

More recently, plants are being engineered to serve as biofactories for the production of edible vaccines. Several food plants, including banana, potato, lettuce, and carrots, can now be modified to induce immunological protection against hepatitis B virus.

One of the major achievements of plant genetic engineering is the "construction" of "golden rice," a rice cultivar with improved nutritional value. Rice, the main staple food in many populous countries, suffers from two major nutritional deficiencies, those of vitamin A and iron. Vitamin A deficiency affects some 400 million people worldwide, leaving them vulnerable to infections and prone to dwarfism and blindness, and iron deficiency leads to anemia. It is estimated that most preschool children and pregnant women in developing countries, and at least 30%–40% in industrialized countries, are iron-deficient. The bioengineering involved in fighting these deficiencies provides an excellent example of the sophistication and power of modern techniques.

The engineering of "golden rice" involved the introduction of several genes of different origins into the plant genome (Fig. 17.3A). Two genes required for the biosynthesis of β-carotene (the precursor to vitamin A) were taken from daffodil (*Narcissus pseudonarcissus*) and from the soil bacterium *Erwinia uredovora*. The end-product of the engineered pathway is the red pigment lycopene, which is further processed to the desired product β-carotene (vitamin A) by endogenous plant enzymes. This also gives the rice the distinctive yellow color for which it is named (Fig. 17.3B).

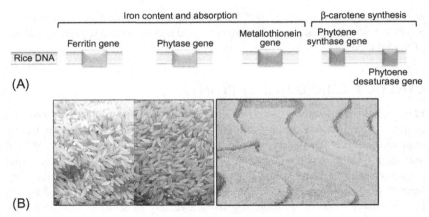

FIG. 17.3 Golden rice. The increased β-carotene content gives the *bright yellow color* to the grains. (A) Genes and their origins introduced into rice to create rice of high nutritional quality. (B) Wild-type and golden-rice grains photographed side by side (from www.goldenrice.org). (C) Fields of golden rice. *((A) Adapted from Garland 2016, Fig. 2 in Box 5.8; (B) Courtesy Golden Rice Humanitarian Board www.goldenrice.org.)*

The iron supplementation project involved introducing three genes into rice plants (Fig. 17.3A). The engineering work was complemented by standard breeding practices that collectively resulted in this "wonder food."

There is controversy concerning GE plants, and it centers around two main issues: the safety of these crops for human consumption and the fear that such artificially created plants may disturb the delicate ecological balance in nature. While the first issue seems to be resolved in favor of GE food, the second is still of concern and a matter of dispute.

GENE THERAPY

Many human disease states can be traced to the absence, or malfunctioning, of one or more genes. Attempts to relieve such diseases by DNA modification are referred to as **gene therapy**. Addition of a functional gene, rather than the protein itself, is preferred because proteins are quickly degraded, whereas a properly integrated gene will continue to be expressed. There are two major approaches:

i. In **gene addition therapy** a normal gene may be inserted into a random (nonspecific) location within the genome to supplement a nonfunctional gene; the nonfunctional gene stays in the genome. Examples might be the use of tumor suppressor genes in the treatment of a malignancy, or immuno-stimulatory genes to treat an infectious disease.

ii. In **gene replacement therapy**, a mutated gene can be swapped for a normal gene through homologous recombination. Exchanges of stretches of nucleotide sequences are broadly defined as **DNA recombination**. When the exchange requires regions of sequence homology between the exchanging partner DNA duplexes, we speak of homologous recombination.

Gene therapy has proved to be a complex process involving several steps in the production of the vector that carries the gene of interest and its introduction into a cell. Once the vector delivers the transgene into the cell, the gene must travel through the cytoplasm and enter the nucleus. Once in the nucleus, the transgene must be stably integrated into the genome: only integrated gene copies can be consistently replicated each time the genome is replicated. Finally, proper, regulated transgene expression must be achieved, which is not a trivial task as the majority of vectors will insert the gene they carry at random locations. Two problems arise from random insertion: (i) in most cases, the gene will end up in a chromosomal environment that does not permit its transcription; and (ii) the gene may land within other genes or their regulatory sequences, which will lead to the inactivation of these host genes, some of which may be essential. Existing methods are constantly being improved or entirely new approaches are created to overcome these problems.

Since the anionic nature of DNA hinders its transfer across the membrane, the focus of an efficient gene therapy is on the vector-mediated delivery. There are two types of vectors: viral and nonviral vectors.

Viral vectors. Viruses possess the natural ability to enter the cell and nucleus efficiently. Vectors based on viruses must be replication-incompetent, i.e., they must carry only the essential elements necessary for transgene packaging and expression. To this end, genes necessary for viral particle production are removed from the vector and replaced with the gene of interest. The overall **transfection efficiency** depends on the efficiency of the steps involved in viral uptake, entry into the nucleus, and escape from degradation. Viruses containing a dsDNA genome have, in general, higher transfection efficiency than ssDNA and RNA viruses; the latter need to first convert their genomes into dsDNA.

Nonviral vectors. Nonviral vectors have been developed in order to overcome the immunogenicity and the safety issues linked to the use of viral vectors. Such vectors, however, have an impaired ability to enter the nucleus; thus far, this feature has limited their use to proliferating cells only, to make use of the availability of stages in proliferation where the genome is not contained within a nuclear envelope.

Alternatively, therapeutic DNA can be directly introduced into target cells. The use of liposomes, artificial lipid spheres with an aqueous core containing the DNA, is another option. In still another approach, the DNA is chemically linked to molecules that are capable of binding to special cell-surface receptors or of facilitating nuclear transfer. Prominent among nonviral vectors are human artificial chromosomes, with practically unlimited capacity and stability of gene expression, and no immunogenicity.

The administration of gene therapy to humans has been and remains controversial. There has been growing skepticism about the efficiency and safety of the procedures. The public became wary after an 18-year-old boy died from multiple organ failure four days after treatment, and several patients in a Paris-based clinical trial developed leukemia. It became painfully clear that treatment is risky, although in many cases successful and life saving. That "gene therapy deserves a fresh chance" (*Nature* 461, 2009, 1173) has been substantiated by recently reported cases in the successful treatment of numerous diseases such as immune deficiencies, hereditary blindness, hemophilia, β-thalassemia, Parkinson's disease, and last, but not least, cancer. A significant improvement in the treatment has been the removal of patient's own cells (blood stem cells, immune cells) that are then treated in culture with a virus genetically altered to recognize a protein that is specific for the surfaces of cancer cells. The treated cells are then returned to the patients. Because these are recognized as "self" cells, there is no immunological reaction. In a clinical trial with 59 leukemia patients, 26 experienced complete remission.

A CRISPR REVOLUTION?

It seems likely that much of the effort that has been put into perfecting gene therapy by traditional recombinant methods shall soon be superseded. As described in Chapter 14, the new CRISPR techniques allow modification, removal, or addition of genes at any desired location in the genome. Problems of inappropriate location of modifications intended for therapy should therefore be eliminated. Modification can take place either in vivo or in vitro, using the patient's own cells, removed from the body, modified, and then reintroduced. As is sometimes the case, science seems to have "leapfrogged" over a seemingly intractable problem! As long as only somatic cells are involved, concerns over the technique should not be consequential. Already there has been success in animal studies.

CLONING OF WHOLE ANIMALS

Although it is not, strictly speaking, a recombinant DNA technique, it seems appropriate to conclude this section on applications of modern gene technology with a brief description of whole-animal cloning. A clone of an animal will, by definition, be an exact copy of that animal, and must, if our current understanding is correct, carry exactly the same genetic information.

Such clones can be produced by nuclear transfer. Recall from Chapter 13 that this method involves taking a cell nucleus from a donor organism and injecting it into an enucleated egg. This hybrid egg, containing a nucleus (read DNA) from the organism to be cloned and the egg's own cytoplasm, is then fertilized with sperm from the donor and allowed to develop as an embryo, which is then transplanted into the uterus of a foster mother. This will develop into a fetus carrying *only* DNA from the donor; all the natural possibilities of genetic exchange in sexual reproduction have been bypassed. The newly born animal is now an exact genetic copy of the donor organism.

The early experiments with amphibian cloning were fairly easy because frog eggs are large, but manipulating the tiny eggs of mammals proved much more difficult. Nevertheless, researchers were able to successfully clone a number of mammals, from mice to sheep. In 1996, the sheep "Dolly" was cloned from an epithelial cell of an adult donor, a major breakthrough that attracted public attention. In 1997, a transgenic lamb "Polly" was cloned from cells engineered with a couple of genes, thus for the first time combining genetic modification with cloning; the lambs produced a new protein of human origin. An amazing achievement reported in 2002 was the modification of cow genomes so that the antibodies the transgenic animals produce are of human origin. The project involved the creation of microchromosomes that carry the large genome regions that encode the immunoglobulins; these were then successfully introduced into cells and transmitted through many cell divisions. The endogenous bovine genes were inactivated to ensure the purity of the human proteins produced; in addition, a cattle gene that may be involved in prion production was deleted to further ensure the safety of the product. The cloned calves continued to produce human immunoglobulins.

2006 witnessed the engineering of a pig to produce omega-3 fatty acids through the insertion into its genome of a roundworm gene. In addition to using pigs for production of pharmaceuticals and proteins for use in clinical practice, pigs were found of great promise for creating models for human genetic diseases. Prior experiments using mice revealed that the rather large differences between mice and humans limit the value of mouse models. Pigs now serve as effective models of cystic fibrosis and Alzheimer's disease. Finally, pig organs are used for transplantation into human patients. Genetically modified pigs are being created to decrease the immunological rejection of transplanted organs.

At first glance, the cloning of domestic animals and pets may seem an attractive alternative to conventional breeding, but it has not yet proved so commercially. Cloning of an animal is a time-consuming and expensive business; often hundreds of trials are required before success. But the wide range of possibilities and the ever-evolving techniques will contribute to furthering the use of cloned animals.

JURASSIC PARK OR DEEXTINCTION

In 1993, film director Steven Spielberg created the movie Jurassic park, describing a fictional, disastrous attempt to create a theme park inhabited by cloned dinosaurs. The public was fascinated and the movie was tremendously successful. Does the idea behind this fantastic plot have any connection to reality?

The answer is yes. Science has advanced so much that it is in fact now possible to conceive projects to bring extinct species back to life, or to save species at the brink of extinction. Even more realistic is the idea to genetically modify closely related living species so that they possess only some of the features of their long-extinct relatives, features that will make them adapted to special present day environments on Earth. The best-known example of this is the "woolly mammoth" (*Mammuthus primigenius*) project. The Harvard geneticist

George Church and his team are working toward creating elephants that have mammoth-derived adaptations to cold climate, with the idea of repopulating the Siberian tundra with these transgenic elephants. Returning these mammoth-like creatures to the tundra would help revive the ancient grassland there, which, in turn, is expected to prevent the melting of Siberian permafrost.

The success of such an ambitious project depends on many steps, some of which are already in place. First, we now have the sequence of the entire woolly mammoth genome. Hairs from two individuals naturally preserved in the Siberian permafrost allowed the reconstruction of the entire sequence and the identification of genes responsible for some of the morphological features needed for survival under cold climates. These include hemoglobin genes, genes encoding the accumulation of subcutaneous fat, genes for long hair and ear size. Fourteen of these genes have been introduced by Church's team into the genome of the modern elephant, and the effort now (2015) focuses on deriving tissues or stem cells from the genetically modified cells in culture. The necessary technological advances are already available to bring the project to the successful end.

There is no doubt that with time more and more projects like this one will be developed, to widen the range of existing animal species. Mankind has long tried to do this through selective breeding. Now, for the first time, we seem to be on the verge of actually controlling evolution. Will it be to the benefit of humanity?

EPILOGUE

The recombinant DNA techniques developed by molecular biologists in just the last few decades have had far-reaching consequences in fields as diverse as forensics, medicine, and agriculture. These in turn have spawned a multitude of high-tech industries. Like many of the scientific advances that have changed our lives, research that began in the ivory towers has come to Main Street.

FURTHER READING

Books and Reviews

Carmichael, L.E., 2013. Gene Therapy. ABDO Publishing Company, Edina, MN. Covers broad range of topics, such as the human genome project, hunting for disease genes, and recombinant DNA technologies used to correct defective genes.

Church, G., 2013. Please reanimate. Sci. Am. 309, 12. Argues that reanimation is about leveraging the best of ancient and synthetic DNA, with the goal to adapt existing ecosystems to radical modern environmental changes, such as global warming, and possibly to reverse those changes.

Glick, B.R., Pasternak, J.J., Patten, C.L., 2010. Molecular Biotechnology: Principles and Applications of Recombinant DNA, fourth ed. ASM Press, Washington, DC. A textbook describing the state-of-the-art technology used currently in recombinant DNA manipulation.

Grierson, D. (Ed.), 2013. Plant Genetic Engineering. Blackie Academic & Professional, Glasgow. Reviews the basic science that underpins plant biotechnology and shows how this knowledge is being used in directed plant breeding. Highlights the fundamental understanding of plant physiology, biochemistry, and cell and molecular biology for the successful genetic engineering of plants.

Herzog, R.W., Zolotukhin, S. (Eds.), 2010. A Guide to Human Gene Therapy. World Scientific Publishing, Hackensack, NJ. Leading experts cover topics on viral and non-viral gene transfer systems and treatment of hematological diseases and disorders of the central nervous system.

Kumar, A., Garg, N. (Eds.), 2005. Genetic Engineering. Nova Science, New York. A textbook that covers comprehensively topics of genetic engineering for the graduate, postgraduate students and young research scholars of biological sciences.

Pena, S.D.J., Chakraborty, R., Epplen, J.T., Jeffreys, A.J. (Eds.), 2013. DNA Fingerprinting: State of the Science. Springer, Basel, AG, Base. Covers a broad range of subjects starting with the basic aspects of the genomic organization of tandemly repeated DNA sequences and then describing the applications of DNA fingerprinting to the study of human populations, microorganisms, plants and animals.

Sambrook, J.B., Russell, D., 2001. Molecular Cloning: A Laboratory Manual, third ed. Cold Spring Harbor Laboratory Press, Cold Spring Harbor, NY. A set of protocols covering various aspect of cloning; widely used in the laboratory.

Classic Research Papers

Campbell, K.H., McWhir, J., Ritchie, W.A., Wilmut, I., 1996. Sheep cloned by nuclear transfer from a cultured cell line. Nature 380, 64–66. The first report for successful nuclear transfer from an established cell line into enucleated oocytes, resulting in the birth of lambs.

Beyer, P., Al-Babili, S., Ye, X., Lucca, P., Schaub, P., Welsch, R., Potrykus, I., 2002. Golden Rice: introducing the beta-carotene biosynthesis pathway into rice endosperm by genetic engineering to defeat vitamin A deficiency. J. Nutr. 132, 506S–510S. Describes the introduction, in a single DNA transformation act, of several genes from different plant origins; the resulting rice plants produced provitamin A in the endosperm.

Goeddel, D.V., Kleid, D.G., Bolivar, F., Heyneker, H.L., Yansura, D.G., Crea, R., Hirose, T., Kraszewski, A., Itakura, K., Riggs, A.D., 1979. Expression in *Escherichia coli* of chemically synthesized genes for human insulin. Proc. Natl. Acad. Sci. USA 76, 106–110. Describes the cloning of synthetic genes for human insulin A and B chains in a bacterial plasmid, their expression in *E. coli*, and purification from the recombinant bacterial cells.

Jeffreys, A.J., Wilson, V., Thein, S.L., 1985. Individual-specific 'fingerprints' of human DNA. Nature 316, 76–79. Alec Jeffreys and coworkers describe the development of DNA probes capable of simultaneously revealing hypervariability at many loci in the human genome.

Glossary

Adaptor Molecules postulated by Francis Crick as intermediates to match amino acids to DNA triplet nucleotide code. Later identified as tRNAs.

Allostery Intramolecular interactions in which the binding of one ligand to a macromolecule can influence its affinity for others.

Amino group ($-NH_2$) Chemically reactive group with one free valence. In general, amines are derivatives of ammonia (NH_3), wherein one or more hydrogen atoms of are replaced by some substituent group, such as an alkyl (e.g., methyl, CH_3-).

Analytical ultracentrifuge A device for separating macromolecules by sedimentation in a centrifugal field, with observation of the sedimentation process.

Archaebacteria The original name for the presently accepted name Archaea. Archaea were originally thought to inhabit only highly inhospitable conditions characterized by extremes of temperature, pH, or high salinity but have since been found in all types of environments.

Artificial chromosomes High capacity vectors which contains sufficient elements of chromosomes (telomeres, centromeres, replication origin) to both accommodate large DNA inserts and replicate in a particular host, for example, yeast.

Bacteria All microbes except Archaea. Bacteria do not have a nucleus (see Nucleus).

Bacterial conjugation Coupling of bacteria to allow DNA transfer between bacterial cells.

Biochemistry The chemistry of living organisms.

Blunt (flush) ends Ends of DNA fragments produced by restriction enzymes that cut the two strands of a DNA helix exactly opposite of each other.

Bragg angle In X-ray diffraction: angle at which reflections from parallel crystal planes reinforce.

Buffer An aqueous solution consisting of a mixture of a weak acid and its conjugate base, or vice versa. The pH of a buffer solution changes very little when a small amount of strong acid or base is added to it.

Carboxyl group ($-COOH$) Chemically reactive group with one free valence. The basic unit of carboxylic acids, which include compounds such as amino acids and acetic acid.

Catalyst A substance that accelerates a chemical reaction without being modified itself.

cDNA (copy DNA) A double-stranded DNA that has been transcribed from RNA by reverse transcriptase (see Reverse transcriptase).

cDNA library A library of cloned DNAs that corresponds to a collection of the expressed messenger RNAs of a cell type or organism.

Central dogma Name given to Francis Crick's hypothesis that information flows only from DNA to protein, not vice versa.

Chloroplasts Membrane-bound organelles found in plant and algal cells, where photosynthesis takes place. In photosynthesis, the photosynthetic green pigment chlorophyll captures the energy from sunlight to synthesize organic molecules from atmospheric carbon dioxide. Chloroplasts carry out a number of other functions, including synthesis of fatty acids and amino acids; they are also involved in the immune response in plants.

Chromatin The DNA-histone complex which constitutes mitotic and interphase chromosomes (see *Histones*).

Chromatography Collective term for a set of laboratory techniques for the separation of mixtures. The mixture is dissolved in a fluid called the mobile phase, which carries it through a structure holding another material called the stationary phase.

Clone Multiple identical copies of an entity.

Cloning Production of multiple identical copies (clones, see) of an entity; in molecular biology, most often a DNA sequence. In biology, often entire organisms.

Codominance Inheritance where neither allele is dominant, with the individuals in the first generation (F1 hybrid) showing traits from both true-breeding parents.

Codon The sequence of three nucleotide residues in DNA that specifies a particular amino acid residue in the genetic code (see Genetic code).

coimmunoprecipitation (co-IP, pulldown) A useful modification of the IP technique that identifies interacting proteins or protein complexes present in complex samples: by immunoprecipitating one protein member of a complex, additional members of the complex may be captured and then identified by independent methods.

Colloid In chemistry, a mixture in which one substance of tiny dispersed insoluble particles is suspended throughout another substance, usually a liquid.

Column chromatography A low-cost method used to purify individual chemical compounds from mixtures of compounds. A matrix (stationary phase) is poured into a column, a solution of the organic material (mobile phase) is pipetted on top of the matrix, and an eluent is slowly passed through the column. The compounds separate because they are retained by the matrix differently, resulting in their running at different speeds.

Conformation In chemistry: three-dimensional spatial structure of a molecule.

Conservative replication A theoretically possible mechanism of DNA replication, in which both strands of the original mother DNA duplex are copied to give rise to an entirely new daughter duplex; the mother duplex is preserved in the process.

CRISPR-Cas (clusters of regularly interspaced palindromic repeats and associated proteins) Technology that can edit genes in living organisms in any desired way, adding or deleting gene sequences or changing sequences in a predetermined way. Based on the existence of a system in bacteria and *Archaea* that protects against viral infection.

C-terminal end The end of a polypeptide chain which has an unreacted -COOH group.

Cytoplasm The material within the cell, excluding the nucleus (in eukaryotes). It consists of cytosol, the gel-like substance within the cell membrane and organelles, the cellular insoluble substructures.

Degenerate code A feature of the genetic code referring to the usage of several codons to specify one and the same amino acid.

Denaturation A process that leads to a change in the native conformation of a polypeptide or polynucleotide. DNA denaturation involves separation of the two complementary strands; protein denaturation disrupts the native folding of the polypeptide chain without breaking covalent bonds.

Density gradient centrifugation A method for separating macromolecules based on their density. A density gradient is produced in a centrifuge tube, either by sedimentation of a heavy salt or by mixing two solutions of different density. During centrifugation, macromolecules in the solution will migrate to positions where their density matches solution density.

Deoxyribonucleic acid (DNA) A polymer of deoxyribonucleotide residues.

Directional selection Natural selection that leads to fixing, with time, of one of the two extremes of a trait, if this extreme confers higher fitness of the organism.

Disaccharide A sugar molecule formed when two monosaccharides (simple sugars) are joined by glycosidic linkage. The three most common disaccharides are sucrose, lactose, and maltose; they all have 12 carbon atoms, with the general formula $C_{12}H_{22}O_{11}$ and differ in atomic arrangements within the molecule.

Dispersive replication A theoretically possible mechanism of DNA replication, in which both daughter duplexes contain mixed (old and new) pieces of DNA in all DNA strands.

Disruptive selection Natural selection when both extremes of a trait are more beneficial in terms of fitness than intermediate types. Selection will lead to fixing both extremes with time.

DNA fingerprinting (DNA profiling, DNA typing) A forensic technique used to identify individuals by characteristics of their DNA. Only about 0.9% of the DNA, so-called repetitive (repeat) sequences, is different from one individual to another, allowing identifying individuals, as uniquely as classical fingerprints (hence the name for the technique).

DNA gyrase The enzyme that relieves strain while double-stranded DNA is being unwound by helicase during DNA replication.

DNA ligase Any of a class of enzymes that catalyze the covalent connection of DNA molecules.

DNA phenotyping A new DNA-based forensic technique that performs high-speed sequencing of genomic DNA from the crime scene; then variations in DNA sequence that determine phenotypic markers like eye color or facial shape are used to construct a predicted image.

DNA Pol I This bacterial polymerase is involved in some DNA repair mechanisms and in processing of Okazaki fragments (see Okazaki fragments) generated during lagging strand synthesis. This is the enzyme discovered by Arthur Kornberg.

DNA Pol II This bacterial polymerase is involved in some DNA repair mechanisms; also thought of a backup of the major DNA Pol III activity.

DNA Pol III A multiprotein enzyme complex, the primary polymerase involved in DNA replication in bacteria.

DNA recombination Resorting of genetic segments by exchanges of stretches of duplex DNAs.

Domains Portions of a macromolecule that exhibit a folded structure that is distinguishable from other parts of the molecule and give evidence of independent folding. Often, the domains in a protein molecule carry out distinct functions, as, for example, DNA-binding domains.

Dominance Relationship between alleles of one gene, in which the effect on phenotype of one allele masks the contribution of a second allele of the same gene. The first allele is dominant and the second allele is recessive (see Recessive allele).

Electrophoresis Migration of macromolecules in an electric field. It is widely used, in many variants, for separating or analyzing mixtures of proteins or nucleic acids.

Embryonic stem (ES) cell Cell from a eukaryotic embryo that has the potential to differentiate into a variety of somatic cells.

ENCODE (Encyclopedia of DNA Elements) A multilaboratory international project that was started shortly after the completion of the Human Genome Project (see Human Genome Project (HGP)) with the aim to mine the entire human genome sequence for various functional elements (see Functional elements).

Enzyme A protein that has catalytic activity that speeds up a biochemical reaction.

Epigenesis One of two competing early ideas concerning embryonic development. Sates that early in life the embryo is an undifferentiated mass and morphologically discernable features appear later in development.

Epigenetic Refers to changes in the informational content of the genome in a cell that are outside of DNA; not heritable through the sexual reproduction cycle.

Eubacteria The original name for the presently accepted name bacteria. Organisms that lack membrane-bound nucleus and organelles.

Eukaryotes The class of organisms which possess nucleated cells. Distinguished from bacteria and archaea, which lack nuclei and other organelles.

Exon A segment of a protein-coding gene that is retained in the mature mRNA (see *Intron*) and is translated into protein.

Expression vectors Cloning vectors that allow expression of genes within host cells; this requires that they contain the proper regulatory sequences for transcription and translation (see Vector).

Feedback regulation Regulation of metabolism in which the product of a long chain of biochemical reactions inhibits the first enzyme in the chain. Feedback regulation saves energy by inhibiting the production of a substance that is not needed at the moment.

Fibrous proteins A type of proteins usually organized as filaments, shaped like rods or wires. Fibrous proteins function for protection and support, forming connective tissue, tendons, bone matrices, and muscle fiber. Examples include keratin, collagen, elastin, and fibroin.

Frameshift A mutation (deletion or insertion of any number of nucleotides other than a multiple of three) that shifts the reading frame of a gene; frameshift mutations lead to the production of a protein whose sequence is totally changed 3′ to the mutation.

Functional elements Elements of the genome of characteristic functions. These include, but are not limiters to, transcription start sites and promoters, enhancers, nucleosome locations, DNA methylation sites.

Gel electrophoresis Electrophoresis in a gel matrix. Agarose (a polysaccharide polymer material, generally extracted from seaweed) or polyacrylamide is the most commonly used matrix-formers.

Gene A region of DNA that encodes for a protein or a functional RNA molecule; the molecular unit of heredity.

Gene addition therapy A normal gene is inserted into a random location within the genome to supplement a nonfunctional gene; the nonfunctional gene stays in the genome.

Gene replacement therapy A mutated gene can be swapped for a normal gene through homologous recombination (see Homologous recombination).

Gene therapy Therapeutic delivery of nucleic acid fragments into a patient's cells as a drug to treat disease. Addition of a functional gene, instead of the protein itself, is preferred because proteins are quickly degraded, whereas a gene integrated into the genome will continue to be replicated during cell division and expressed cell generation after cell generation.

Genetic code The set of rules that govern the translation of information in nucleic acids (DNA and RNA) into protein sequences.

Genetics The study of genes and their role in inheritance.

Genomic libraries Large sets of recombinant DNA molecules (vectors, see, plus genomic inserts) that contain entire genomes in the form of overlapping sequences. Genomic libraries can be multiplied in bacteria and stored indefinitely.

Genomics At the molecular level, the study of the entire genetic content (genome) of a cell or a virus. It applies DNA-sequencing methods and bioinformatic analysis of the data to understand the structure and function of genomes.

Genus The first part of the binomial system to identify a living organism. For example, *Homo sapiens* (modern-day humans), *Homo* is the name of the genus.

Heterokaryon A cell containing two nuclei of different origin formed initially upon fusion of two different cells (see, e.g., normal somatic cell with a pluripotent cell).

HGP-Write A recent extension to the Human Genome Project (see Human Genome Project) aiming at synthesizing huge stretches of the human genome to be used for scientific and medicinal purposes.

Hierarchical shotgun sequencing approach In sequencing whole genomes, a set of large clones, typically 100–200 kbp each, is generated and then organized (through restriction nuclease mapping). Appropriately chosen clones are then subjected to shotgun sequencing (see Shotgun sequencing). This was the approach used for sequencing of the human genome.

Histones Small, evolutionarily conserved basic proteins that constitute the protein basis of the structure of eukaryotic chromatin.

Homologous recombination Recombination between DNA fragments that requires strong sequence homology between the recombining DNA sequences.

Host organism In biology, a host is an organism that harbors a parasitic or a symbiotic organism, typically providing nourishment and shelter. In recombinant DNA technology, an organism that can reproduce to a large number, thus creating multiple copies of an inserted DNA sequence.

Host restriction A term to describe the ability of some bacteria to restrict the growth of certain bacteriophage. It was explained by the existence of restriction endonucleases and DNA methyl transferases. Methylation protects the host DNA from the endonuclease, which cleaves only the unmethylated sequences of the invading phage.

Human Genome Project (HGP) An international collaborative project launched in 1990 with the ambitious goal of determining the sequence of the entire human genome, and identifying and mapping all human genes. Initial rough draft became available in 2000, with a final draft appearing in 2003.

Hydrolysis Cleavage of a covalent bond with the introduction of a water molecule to carry out the reaction. The hydrogen and hydroxyl from the water molecule are attached to the two fragments produced.

Hydrophilic Water-liking chemical groups.

Hydrophobic Water-avoiding chemical groups.

Immunological reactions Reactions of the immune system aimed at neutralizing harmful invading agents and substances foreign to the body. Involves numerous cellular responses and the production of immunoglobulins, antibody molecules that specifically recognize and bind to the harmful agent.

Immunoprecipitation (IP) A method to purify proteins out of complex biological samples, such as cell or nuclear extracts or bodily fluids that may contain thousands of different protein molecules. The method is based on the ability of antibodies against the protein of interest to form insoluble complexes with that protein.

Incomplete dominance Inheritance where the hybrid resembles neither parent, i.e., the phenotype of the F1 hybrid is intermediate between the phenotypes of the parents.

Induced fit A model of enzyme action which proposes that the enzyme will fit the substrate only if the latter is distorted into a conformation part-way into the reacted form. This makes it easier for the reaction to go to completion.

Induced pluripotent stem cells (iPSC) Differentiated cells that can be reverted to pluripotency by different methods (by expression of certain protein factors or by exposure of cultures to a set of small molecules).

Inducer A low-molecular-weight substance that promotes expression of a bacterial operon by binding to a repressor molecule, causing it to dissociate from the gene promoter.

Intron (aka intervening sequence) Any nucleotide sequence within a gene that does not code for any portion of a polypeptide. The term refers to both the DNA sequence within a gene and the corresponding sequence in the primary RNA transcripts (pre-mRNA, see). Introns are removed by RNA splicing (see Splicing) during maturation of the primary transcript into a functional mRNA molecule.

Isoelectric focusing A technique for separating molecules by differences in their isoelectric points (see Isoelectric point). Usually performed in a gel matrix that contains a gradient of pH: the molecule will stop moving when its isoelectric point equals that of the surrounding gel matrix.

Isoelectric point The pH at which a protein carries zero net charge.

Junk DNA A term formerly applied to noncoding DNA in the genome which was thought to have no useful function. We now know that most or all is transcribed, and thus may function in certain ways, so the term is no longer used.

Keratin A protein in the hair that has a fibrous conformation.

Knockdowns Cells or organisms in which the regulation of a particular gene is modified, usually to decrease its function. The reduction can be achieved through modification of nucleotide sequences or treatment with reagents such as short DNA or RNA oligonucleotides that possess sequence complementary to either gene or an mRNA transcript.

Knockin organisms Organisms in which a modified gene is specifically substituted for the wild-type gene. Knockin also refers to insertion of sequence information not found within the locus. In both cases, these are "targeted" insertions into specific loci.

Knockout organisms Organisms in which a specific gene has been inactivated by inserting an extraneous DNA sequence into them. Knockouts provide the most definitive evidence for the function of a gene.

Lagging strand The DNA strand that is synthesized discontinuously during replication.

Leading strand The DNA strand that is synthesized continuously during replication.

Linked genes Linked genes are found not to segregate independently, an exception to Mendel's second law (see Mendel's second law). Linked genes are located in close proximity on chromosomes.

Meiosis Cell division to produce haploid gametes, each with only one copy of the genome.

Mendel's first law (the law of segregation) The two alleles for each trait separate (segregate) during gamete formation, then unite at random, one from each parent, at fertilization.

Mendel's second law (the law of independent assortment) During gamete formation, the segregation of the alleles of one allelic pair is independent of the segregation of the alleles of another allelic pair. In other words, traits segregate independently; there is no linkage between genes for different traits.

messenger RNA (mRNA) The RNA that is complementary to one strand of a protein-coding DNA gene, and carries this sequence to the ribosome to direct protein synthesis.

Mitochondria (plural for mitochondrion) Membrane-bound organelles found in all eukaryotic organisms. Mitochondria generate most of the cell's supply of adenosine triphosphate (ATP), used as an energy source for biochemical reactions.

Mitosis Cell division in which two double-strand copies of parental DNA are produced and the resulting daughter cells contain a diploid number of chromosomes. It is typical of eukaryotic somatic cells.

Molecular biology The study of biology at the molecular level.

Monoclonal antibody A particular antibody molecule produced by a single B-cell clone. The normal immune response results in the production of a large number of different antibodies (polyclonal antibodies), produced by clones of individual cells that differ in exact organization of their immunoglobulin genes.

Multifactorial inheritance group A group of interacting genes that determine a single trait. Novel phenotypes can emerge from the combined action of the alleles of two genes.

Multipotency Developmental potential of cells capable of giving rise to cell types of a given cell lineage (e.g., hematopoietic cells can differentiate into lymphocytes, monocytes, erythrocytes, and other blood cell types but cannot give rise to, for example, brain cells). Most adult stem cells are multipotent. These include skin stem cells, hematopoietic, and neural stem cells.

Mutation Permanent alteration of the nucleotide sequence of any genetic element. Mutations result from errors during DNA replication or from damage to DNA caused by physical or chemical agents.

Myoglobin A smaller relative of hemoglobin found in muscle tissue that stores oxygen transported by hemoglobin.

Nonviral vectors Developed to overcome safety issues of viral vectors. Some vectors use artificial lipid spheres, liposomes, that contain DNA in their aqueous core. In other cases, DNA molecules are chemically linked to molecules that facilitate their traversing cell or nuclear membranes. Artificial human chromosomes (see Artificial chromosomes) belong to this class too.

N-terminal end The end of a polypeptide chain that carries an unreacted -NH_2 group.

Nuclear magnetic resonance (NMR) A physical phenomenon in which nuclei in a magnetic field absorb and reemit electromagnetic radiation. NMR allows the observation of specific quantum mechanical magnetic properties of the atomic nucleus. NMR is the theoretical basis of NMR spectroscopy, a method to study atomic structures of molecules in solution.

Nucleic acid hybridization A technique in which single-stranded nucleic acids (DNA or RNA) are allowed to interact so that complementary sequences form double helical structures; these are called hybrids since the components they may come from different sources.

Nucleic acids (polynucleotides) Polymers of nucleotide residues. DNA and RNA.

Nucleosome General term for DNA-histone octamer particles that are major repeating units of eukaryotic chromatin (see nucleosome core particle, octasome).

Nucleus A membrane-enclosed organelle found in eukaryotic cells. The nucleus contains most of the cell's genetic material in the form of several long linear DNA molecules bound by proteins to form chromatin (see Chromatin).

Okazaki fragments The short pieces of DNA strand synthesized on the lagging strand (see Lagging strand) template in replication of double-strand DNA.

Operator In bacteria, a genomic region which binds a protein and thereby regulates an operon (see Operon).

Operon In bacteria, a group of contiguous genes of related function and regulated as a group.

Organelle Specialized cellular compartment that has a highly specific function. Examples are mitochondria (see Mitochondria) and chloroplasts (see Chloroplasts). Organelles are membrane-bound and are localized within the cytoplasm.

Overhangs (sticky ends) At a DNA end, extension of one strand beyond the other. Also called sticky ends because they facilitate interaction between DNA fragments that contain the same sticky ends; these ends interact through hydrogen bonding (as in a double-stranded DNA), keeping the fragments in position for ligation by DNA ligases (see DNA ligase).

Paper chromatography An analytical method used to separate colored chemicals or substances. A paper chromatography variant, two-dimensional chromatography, involves using two solvents and rotating the paper 90 degrees in between.

Peptide bond The covalent bond formed by the elimination of a water molecule between the amino group of one amino acid and the carboxyl group of another.

Permease A membrane-bound transport protein instrumental in getting small molecules from outside into a cell.

Phase problem In X-ray diffraction: the problem that intensity of spots on a diffraction pattern does not alone provide sufficient information to solve a molecular structure; the relative phases of radiation scattered in these directions must also be determined. This can be done by preparing crystals with replacement of specific atoms by heavy metal atoms.

Phenocopying A phenomenon where environmental agents cause a change in phenotype that mimics the effects of a mutation in a gene.

Phenotype Morphologically observable characters.

Pitch (of a helix) The distance along the helix axis at which the helical structure repeats.

Plasmids Small, extrachromosomal DNA molecules found in most bacterial cells; they can replicate independently of the cell DNA. Plasmids were the first (and still often used) cloning vectors (see Vector).

Pleiotropy The phenomenon in which one gene may contribute to several, seemingly unrelated, visible characteristics.

Pluripotency The ability to produce all cells of the embryo proper (excluding extraembryonic tissues). Pluripotency is complete, when a cell can form every cell of an organism (e.g., cells of the inner cell mass and embryonic stem cells (see Embryonic stem cell) derived from these cells and grown in culture); partially pluripotent cells can differentiate into limited cell types.

Polymerase Any of a large group of enzymes which catalyze the polymerization of nucleic acid monomers. There are RNA polymerases and DNA polymerases. All add monomers in the 5′–3′ direction. Most, but not all, require a template DNA or RNA.

Polymerase chain reaction (PCR) An in vitro technique for greatly increasing the amount of a desired DNA fragment by repeated cycles of copying strands of denatured DNA templates starting from primer oligonucleotides.

Polymers Giant molecules that are formed by covalent addition of similar units.

Polypeptide Long chains of amino acids covalently linked via peptide bonds (see Peptide bond).

Preformation One of two competing early ideas concerning embryonic development. States that the embryo is a miniature individual that preexists in either the mother's egg or the father's sperm. Later in development, this miniature individual just grown *in utero*.

Pre-mRNA The primary transcript of a gene that contains the entire gene sequence, including introns (see Intron) and exons (see Exon). This primary transcript has to undergo further processing to remove the noncoding DNA sequences in a process known as splicing (see Splicing).

Primary structure The sequence (order) of amino acid residues in a protein molecule.

Primase The enzyme that synthesizes primers (see Primer) on DNA segments to be replicated.

Primer In replication, a short oligonucleotide which base-pairs with the DNA template strand and provides a free 3′-OH group for primer extension by polymerase from this point. Also refers to DNA primers used in PCR (see Polymerase Chain Reaction (PCR)).

Principle of genetic equivalence States that all cells of an adult organism contain the same genetic information and differ only in what portion of that information is being used in a specific cell type. In other words, nuclei from differentiated cells retain all the genetic information needed for the development of an entire organism; they just need to be "reprogrammed" by, for example, being transferred to the environment of an enucleated egg.

Prokaryote Unicellular organism that lacks a membrane-bound nucleus (karyon) and membrane-bound organelles, including mitochondria. Prokaryotes are represented by two domains, Archaea (see Archaebacteria) and bacteria (see Bacteria).

Promoter A genetic element that is essential for the expression of a gene. Promoters usually lie near transcription start sites and bind RNA polymerase and initiation factors.

Proteomics The study of the proteome, ideally, the complete list of proteins and their interactions, in each organism.

Pseudogene A genetic element very similar or identical to a protein gene, but which cannot be expressed for lack of a functional promoter.

Purine base An organic compound that consists of two heterocyclic rings (heterocyclic rings have atoms of at least two different elements as members of the ring). In DNA, adenine and guanine.

Pyrimidine base An organic compound that consists of one heterocyclic ring (heterocyclic rings have atoms of at least two different elements as members of the ring). In DNA, cytosine and thymine; in RNA, uracil.

Quaternary structure The level of protein structure produced by (usually) noncovalent association between protomers. Protomers are units in a multisubunit protein, or in any protein with quaternary structure. Protomers in a single multisubunit protein may be of one type or many.

Recessive allele The allele of a gene whose phenotype remains hidden in the first generation (hybrid) progeny (F1 generation). I then "reappears" in following generations.

Recombinant DNA technology The technology that utilizes recombinant DNA techniques, either in research or in industry.

Renaturation A process that leads to restoring the native conformation of a polypeptide or polynucleotide. The inverse process of denaturation.

Replisome The multiprotein complex which replicates DNA.

Repressor A protein that represses transcription by binding to specific nucleotide sequences, usually located in regulatory regions in or near a gene.

Restriction endonucleases (restrictases) Enzymes that cleave nucleic acids at specific nucleotide sequences.

Reverse transcriptase (RT) Enzyme, first discovered in retroviruses that catalyzes the formation of double-stranded DNA by use of an RNA molecule as template. Also known as **RNA-dpenent DNA polymerase**.

Ribonucleic acid (RNA) A polymer of ribonucleotide residues.

Ribosomes Subcellular particle, composed of RNA and proteins, that are the sites of protein synthesis.

RNA Tie Club An informal scientific club founded in 1954 by James Watson and George Gamow to discuss issues relating to the genetic code and the way it is read by the cell. Consisted of 20 members, each representing 1 amino acid, and 4 additional "honorary" members representing the 4 nucleotides.

Satellite DNA Some small DNA segments will have density quite different from bulk DNA. Therefore, they will band separately (satellite bands) in a density gradient in the ultracentrifuge.

SDS-gel electrophoresis Gel electrophoresis of proteins carried out in the presence of the nonionic detergent sodium dodecyl sulfate (SDS). Under these conditions, proteins unfold and can be separated according to chain length.

Secondary structure Portions of regular, repeating folded structure in a protein molecule. The α-helix and β-sheet structures are the most important, but certain types of turns of the chain between α-helices and β-sheets may be included, as well as other helix types.

Sedimentation The falling of molecules or particles through fluid in gravity or in a centrifugal field.

Semiconservative replication A mechanism of DNA replication, in which the original mother duplex is split, and a complementary copy is made of each strand. Thus, each daughter DNA duplex contains an old strand and a newly synthesized strand. This is the actual mechanism of DNA replication.

Sexual reproduction A form of reproduction where two morphologically distinct types of specialized reproductive cells (gametes) fuse together. The female gamete is a large ovum (or egg) and that of the male is a smaller sperm. Each gamete contains half the number of chromosomes of normal cells.

Shotgun sequencing Strategy of sequencing large genomes. Involves creation of a large series of recombinant clones, the determination of their nucleotide sequences, and alignment of clones that contain overlapping sequences. Computational methods are used to reconstruct the sequence of the entire genome under investigation.

Signaling molecules (morphogens) Act over distances of a few to several dozen cell diameters. Diffuse through the tissues of an embryo, setting up concentration gradients. This graded distribution of morphogens within the embryo exposes cells to different morphogen concentrations, activating different transcriptional programs leading to different cells fates.

Single-gene inheritance group A group of genes in the genome where a morphologically observable trait is encoded by a single gene.

Site-directed mutagenesis A technique for the production of sequence modifications (mutants) at specific desired sites in a cloned portion of the genome. Such will be heritable.

Sodium dodecyl sulfate (SDS) A nonionic detergent often used to solubilize proteins.

Species The second part of the binomial system to identify a living organism. For example, *Homo sapiens* (modern-day humans), *sapiens* is the name of the species.

Splicing The nuclear processing of premessenger RNA molecules that removes introns and reconnects exons in the uninterrupted coding sequence of mature mRNA.

Spontaneous generation Early beliefs that living organisms can form spontaneously from nonliving matter, without descent from similar organisms. Certain forms such as fleas were believed to arise from dust; maggots were thought to arise from dead flesh.

Stabilizing selection Natural selection that acts against changes of an intermediate trait that confers highest fitness of an organism. Also known as negative (or purifying) selection because it occurs through selective removal of gene alleles (and thus, traits) that are deleterious.

Stem cell An undifferentiated eukaryotic cell which may differentiate to produce cells with differing capabilities.

Substrate In chemistry, the molecule that undergoes a chemical reaction to generate a product. In biochemistry, an enzyme substrate is the material upon which an enzyme acts.

Sucrose An enzyme that catalyzes the hydrolysis of the disaccharide sucrose into the monosaccharides glucose plus fructose.

Temperate phage Bacteriophage capable of two life styles: lytic, where they reproduce and lyse the bacterium, and lysogenic, when then integrate their genome into the bacterial genome. The integrated phage genome replicates together with the bacterial DNA and may stay "hidden" for many generations until some stress event leads to excision of its genome from the chromosome and transition back to a lytic state.

Tertiary structure The three-dimensional folded structure of a protein or RNA molecule.

Tetranucleotide hypothesis The early idea that DNA is composed of aggregates of "tetranucleotides," each containing one each of the four bases. Proven to be incorrect.

Totipotency With respect to cells, the ability to differentiate into every cell type of an organism. Fertilized eggs (zygotes, see) are totipotent.

Transcription The "reading" (copying) of a DNA template strand into an RNA sequence.

Transduction Transfer of DNA sequences from one bacterium to another via infection with temperate phage (see Temperate phage).

Transfection efficiency The overall efficiency of action of viral vectors; depends on the efficiency of the individual steps involved in viral uptake, entry into the nucleus, and escape from degradation. Viruses with double-stranded DNA genomes are more stable than those having single-stranded DNA or RNA as their genomic material.

Transformation Any change in an organism's genome caused by introduction of foreign DNA. The term malignant transformation is used to indicate progression to a cancerous state in animal cells, independently of the mechanism.

Transgenic organisms Refer to organisms in which a gene sequence that has been isolated from one organism is introduced into a different organism. Usually used to introduce human disease genes into strains of laboratory mice to study the function or pathology involved with that particular gene.

Translation The production of a specific polypeptide (protein) in response to a specific mRNA. Translation occurs on the ribosome (see Ribosomes) mediated by transfer RNAs.

Transposon (transposable element) A DNA element that can move (transpose itself) from one place to another in the genome.

Transposon (transposable element) A DNA element that can move (transpose itself) from one place to another in the genome.

Two-dimensional gel electrophoresis Gel electrophoresis that is carried out sequentially in two dimensions. The first dimension for protein separation may involve, for example, separation by isoelectric focusing, followed by SDS gel analysis in the second dimension.

Unipotent cells Terminally differentiated cells and committed progenitor cells, such as erythroblast (immature erythrocytes).

Unit cell The unit of volume in a crystal which repeats over and over to define the whole crystal.

Unit of heredity A term in classical genetics, today's equivalent is gene.

Vector A DNA used for transmitting other DNA molecules into cells. Plasmids, bacteriophage, and artificial chromosomes all are used as vectors in recombinant DNA technology (see Recombinant DNA technology).

Viral vectors Viruses are used as vectors (see Vector) to introduce foreign DNA into human cells since they possess the natural ability to enter the cell and the nucleus efficiently. Viral vectors are genetically manipulated so that they cannot replicate: the genes necessary for viral particle production are removed from the vector and replaced with the gene of therapeutic interest.

Vitalism A philosophical doctrine which assets that there is a fundamental difference in the nature of living versus nonliving matter; the soul has a nonmaterial nature.

Whole-genome shotgun approach In sequencing whole genomes, the entire genome is cloned as a series of recombinants, and each clone is sequenced. Computational methods are then used to reconstruct the entire genome, through alignment of clones that contain overlapping sequences.

X-ray diffraction A technique used for determining the atomic structure of a crystal, in which the crystalline atoms cause a beam of incident X-rays to diffract into many directions. By measuring the angles and intensities of these diffracted beams, a three-dimensional picture of the density of electrons within the crystal can be deduced. The electron density can be further used to determine the mean positions of the atoms in the crystal.

Zygote A diploid cell formed during fertilization, through unification of haploid gametes (eggs and sperm). Gives rise during differentiation to numerous distinct cells types in the adult body.

α-helix One of the two main types of secondary structures in polypeptide chains. The helix is right-handed, with a rise of 0.15 nm per amino acid residue and 3.6 residues per turn. The helix is stabilized by hydrogen bonds between the carbonyl oxygen of each residue and the amide hydrogen of the forth downstream residue (counted from N- to C-terminus).

β-sheet One of the two main types of secondary structures in polypeptide chains. Two or more chains lie side-by side, parallel or antiparallel, connected by hydrogen bonds between amide hydrogens and carbonyl oxygens on adjacent chains.

Appendix

Nobel Prize Laureates That Have Contributed to the Development of Molecular Biology

The Nobel Prize in Physiology or Medicine has been awarded 107 times to 211 Nobel Laureates between 1901 and 2016. The Nobel Prize in Chemistry has been awarded 108 times to 175 Nobel Laureates between 1901 and 2016.

Physiology or Medicine		
Year	Laureates	Awarded for
2016	Yoshinori Ohsumi	"his discoveries of mechanisms for autophagy"
2015	William C. Campbell and Satoshi Ōmura	"their discoveries concerning a novel therapy against infections caused by roundworm parasites"
2013	James E. Rothman, Randy W. Schekman, and Thomas C. Südhof	"their discoveries of machinery regulating vesicle traffic, a major transport system in our cells"
2012	Sir John B. Gurdon and Shinya Yamanaka	"the discovery that mature cells can be reprogrammed to become pluripotent"
2010	Robert G. Edwards	"for the development of in vitro fertilization"
2009	Elizabeth H. Blackburn and Carol W. Greider	"the discovery of how chromosomes are protected by telomeres and the enzyme telomerase"
2008	Harald zur Hausen	"his discovery of human papilloma viruses causing cervical cancer"
	Françoise Barré-Sinoussi and Luc Montagnier	"their discovery of human immunodeficiency virus"

Continued

Physiology or Medicine—cont'd

Year	Laureates	Awarded for
2007	Mario R. Capecchi, Sir Martin J. Evans, and Oliver Smithies	"their discoveries of principles for introducing specific gene modifications in mice by the use of embryonic stem cells"
2006	Andrew Z. Fire and Craig C. Mello	"their discovery of RNA interference—gene silencing by double-stranded RNA"
2005	Barry J. Marshall and J. Robin Warren	"their discovery of the bacterium *Helicobacter pylori* and its role in gastritis and peptic ulcer disease"
2004	Richard Axel and Linda B. Buck	"their discoveries of odorant receptors and the organization of the olfactory system"
2003	Paul C. Lauterbur and Sir Peter Mansfield	"their discoveries concerning magnetic resonance imaging"
2002	Sydney Brenner, H. Robert Horvitz, and John E. Sulston	"their discoveries concerning genetic regulation of organ development and programmed cell death"
2001	Leland H. Hartwell and Tim Hunt	"their discoveries of key regulators of the cell cycle"
1999	Günter Blobel	"the discovery that proteins have intrinsic signals that govern their transport and localization in the cell"
1997	Stanley B. Prusiner	"his discovery of Prions—a new biological principle of infection"
1995	Edward B. Lewis, Christiane Nüsslein-Volhard, and Eric F. Wieschaus	"their discoveries concerning the genetic control of early embryonic development"
1993	Richard J. Roberts and Phillip A. Sharp	"their discoveries of split genes"
1992	Edmond H. Fischer and Edwin G. Krebs	"their discoveries concerning reversible protein phosphorylation as a biological regulatory mechanism"
1989	J. Michael Bishop and Harold E. Varmus	"their discovery of the cellular origin of retroviral oncogenes"
1987	Susumu Tonegawa	"his discovery of the genetic principle for generation of antibody diversity"
1984	Niels K. Jerne, Georges J.F. Köhler, and César Milstein	"theories concerning the specificity in development and control of the immune system and the discovery of the principle for production of monoclonal antibodies"
1983	Barbara McClintock	"her discovery of mobile genetic elements"

Physiology or Medicine—cont'd

Year	Laureates	Awarded for
1978	Werner Arber, Daniel Nathans, and Hamilton O. Smith	"the discovery of restriction enzymes and their application to problems of molecular genetics"
1975	David Baltimore, Renato Dulbecco, and Howard Martin Temin	"their discoveries concerning the interaction between tumour viruses and the genetic material of the cell"
1974	Albert Claude, Christian de Duve and George E. Palade	"their discoveries concerning the structural and functional organization of the cell"
1972	Gerald M. Edelman and Rodney R. Porter	"their discoveries concerning the chemical structure of antibodies"
1969	Max Delbrück, Alfred D. Hershey, and Salvador E. Luria	"their discoveries concerning the replication mechanism and the genetic structure of viruses"
1968	Robert W. Holley, Har Gobind Khorana, and Marshall W. Nirenberg	"their interpretation of the genetic code and its function in protein synthesis"
1966	Peyton Rous	"his discovery of tumour-inducing viruses"
1965	François Jacob, André Lwoff, and Jacques Monod	"their discoveries concerning genetic control of enzyme and virus synthesis"
1962	Francis Harry Compton Crick, James Dewey Watson, and Maurice Hugh Frederick Wilkins	"their discoveries concerning the molecular structure of nucleic acids and its significance for information transfer in living material"
1959	Severo Ochoa and Arthur Kornberg	"their discovery of the mechanisms in the biological synthesis of ribonucleic acid and deoxyribonucleic acid"
1958	George Wells Beadle and Edward Lawrie Tatum Joshua Lederberg	"their discovery that genes act by regulating definite chemical events" "his discoveries concerning genetic recombination and the organization of the genetic material of bacteria"
1952	Selman Abraham Waksman	"his discovery of streptomycin, the first antibiotic effective against tuberculosis"
1946	Hermann Joseph Muller	"the discovery of the production of mutations by means of X-ray irradiation"
1945	Sir Alexander Fleming, Ernst Boris Chain, and Sir Howard Walter Florey	"the discovery of penicillin and its curative effect in various infectious diseases"
1933	Thomas Hunt Morgan	"his discoveries concerning the role played by the chromosome in heredity"

Continued

Chemistry		
2016	Jean-Pierre Sauvage, Sir J. Fraser Stoddart, and Bernard L. Feringa	"the design and synthesis of molecular machines"
2015	Tomas Lindahl, Paul Modrich, and Aziz Sancar	"mechanistic studies of DNA repair"
2014	Eric Betzig, Stefan W. Hell, and William E. Moerner	"the development of super-resolved fluorescence microscopy"
2013	Martin Karplus, Michael Levitt, and Arieh Warshel	"the development of multiscale models for complex chemical systems"
2012	Robert J. Lefkowitz and Brian K. Kobilka	"studies of G-protein-coupled receptors"
2009	Venkatraman Ramakrishnan, Thomas A. Steitz, and Ada E. Yonath	"studies of the structure and function of the ribosome"
2006	Roger D. Kornberg	"his studies of the molecular basis of eukaryotic transcription"
2004	Aaron Ciechanover, Avram Hershko, and Irwin Rose	"the discovery of ubiquitin-mediated protein degradation"
2003	Peter Agre Roderick MacKinnon	"the discovery of water channels" "structural and mechanistic studies of ion channels"
2002	John B. Fenn and Koichi Tanaka Kurt Wüthrich	"their development of soft desorption ionisation methods for mass spectrometric analyses of biological macromolecules" "his development of nuclear magnetic resonance spectroscopy for determining the three-dimensional structure of biological macromolecules in solution"
1997	Paul D. Boyer and John E. Walker Jens C. Skou	"their elucidation of the enzymatic mechanism underlying the synthesis of adenosine triphosphate (ATP)" "the first discovery of an ion-transporting enzyme, Na^+, K^+-ATPase"
1993	Kary B. Mullis Michael Smith	"his invention of the polymerase chain reaction (PCR) method" "his fundamental contributions to the establishment of oligonucleotide-based, site-directed mutagenesis and its development for protein studies"
1991	Richard R. Ernst	"his contributions to the development of the methodology of high resolution nuclear magnetic resonance (NMR) spectroscopy"

Chemistry—cont'd		
1989	Sidney Altman and Thomas R. Cech	"their discovery of catalytic properties of RNA"
1982	Aaron Klug	"his development of crystallographic electron microscopy and his structural elucidation of biologically important nucleic acid-protein complexes"
1980	Paul Berg	"his fundamental studies of the biochemistry of nucleic acids, with particular regard to recombinant-DNA"
	Walter Gilbert and Frederick Sanger	"their contributions concerning the determination of base sequences in nucleic acids"
1974	Paul J. Flory	"his fundamental achievements, both theoretical and experimental, in the physical chemistry of the macromolecules"
1972	Christian B. Anfinsen	"his work on ribonuclease, especially concerning the connection between the amino acid sequence and the biologically active conformation"
	Stanford Moore and William H. Stein	"their contribution to the understanding of the connection between chemical structure and catalytic activity of the active centre of the ribonuclease molecule"
1964	Dorothy Crowfoot Hodgkin	"her determinations by X-ray techniques of the structures of important biochemical substances"
1962	Max Ferdinand Perutz and John Cowdery Kendrew	"their studies of the structures of globular proteins"
1958	Frederick Sanger	"his work on the structure of proteins, especially that of insulin"
1957	Lord (Alexander R.) Todd	"his work on nucleotides and nucleotide co-enzymes"
1955	Vincent du Vigneaud	"his work on biochemically important sulphur compounds, especially for the first synthesis of a polypeptide hormone"
1954	Linus Carl Pauling	"his research into the nature of the chemical bond and its application to the elucidation of the structure of complex substances"
1948	Arne Wilhelm Kaurin Tiselius	"his research on electrophoresis and adsorption analysis, especially for his discoveries concerning the complex nature of the serum proteins"

Continued

Chemistry—cont'd

1946	James Batcheller Sumner	"his discovery that enzymes can be crystallized"
	John Howard Northrop and Wendell Meredith Stanley	"their preparation of enzymes and virus proteins in a pure form"
1935	Frédéric Joliot and Irène Joliot-Curie	"in recognition of their synthesis of new radioactive elements"
1926	The (Theodor) Svedberg	"his work on disperse systems"
1911	Marie Curie, born Sklodowska	"in recognition of her services to the advancement of chemistry by the discovery of the elements radium and polonium, by the isolation of radium and the study of the nature and compounds of this remarkable element"
1908	Ernest Rutherford	"his investigations into the disintegration of the elements, and the chemistry of radioactive substances"

Index

Note: Page numbers followed by f indicate figures, t indicate tables, and b indicate boxes.

Protein *(Continued)*
  developments, 10*f*
  electrophoresis, 23–24
  elemental composition, 9–10
  enzymes
    catalyst, 10–11
    crystallization, 11
    functions, 12*t*
    hydrolytic splitting of sucrose, 11, 12*f*
    induced-fit model, 13, 15*f*
    lock-and-key model, 13, 15*f*
    metabolic pathway regulation, 12–13, 14*f*
    nuclease, 11–12
  Fischer-Hofmeister hypothesis, 19–20
  gluten, 9
  hemoglobin, 9, 25
  immunological methods, 26*b*
  insulin, 29–30
  molecular structure, 11
  nuclear magnetic resonance, 43
  one-dimensional amino acid sequence, 89
  posttranslational modifications, 176
  sedimentation, 21–22
  splicing, 176
  trans-splicing, 176
  x-ray diffraction
    crystals, 35–37
    domains, 42–43
    globular protein, 40–43
    α-helix and β-sheet, 39*f*
    keratin, 33, 38*f*
    polymer fiber, 35
    principle, 34, 34*f*
Proteomics, 43
Pseudogenes, 177–178

R

Radical "adaptor" hypothesis, 89
Recombinant DNA technology
  applications, 195–196
  cloning vectors, 152*b*
  CRISPR techniques, 200
  DNA cloning, 150, 151*f*
  DNA ligases, 155
  DNA profiling, 193–194
  gene therapy, 198–200
  genetically modified crops, 149–150
  HIV infection detection, 196–197
  host-controlled variation, 154
  mutation, 149
  natural DNA modification, 155–156

PCR amplification reaction, 157, 158*f*
peptide-coding gene, 157
phage DNA degradation, 154
plant genetic engineering, 197–198
plasmids, 156
restriction endonucleases, 150–155
  cleavages, 154, 155*f*
  discovery, 150
  DNA protection, 154
  host restriction, 154
  restriction process, 154
site-directed mutagenesis, 157, 159*f*
transgenic organisms, 159–160
vaccine production, hepatitis B, 195–196
whole-animal cloning, 200–201
woolly mammoth project, 201–202
Redi, F, 3–4
Replication of DNA
  bidirectional, 84–85, 85*f*
  chemistry, 80, 82*f*
  conservative replication, 79, 80*f*
  core proteins, 84*f*
  density gradient centrifugation, 79–80
  dispersive replication, 79, 80*f*
  E. coli mutant, 81
  Kornberg enzyme, 81
  leading strand, 82–83
  Meselson Stahl experiment, 79–80, 81*f*
  polymerization reaction, 80–83
  replisome, 83, 84*f*
  semiconservative replication, 79, 80*f*
Replisome, 83, 84*f*
Repressor, 111
Restriction fragment length polymorphism
  (RFLP), 194
Reverse transcriptase polymerase chain
  reaction (RT-PCR), 196–197
Ribonucleic acid (RNA)
  editing, 176
  genetic information transfer
    messenger RNA, 91–92
    ribosomal RNA, 91
  nucleosides, 57–58, 59*f*
  polymerase, 107
Ribosomal RNA, 91
Ribosomes, 104–105

S

Saccharomyces cerevisiae, 67
Satellite DNA, 131
Schizosaccharomyces pombe, 67

Printed in the United States
By Bookmasters